Biobanks

AF736426

In recent years, a number of large population-based biobanks – genetic databases that combine genetic information derived from blood samples with personal data about environment, medical history, lifestyle or genealogy – have been set up in order to study the interface between disease and genetic and environmental factors. Unsurprisingly, these studies have sparked a good deal of controversy and the ethical and social implications have been widely debated.

Biobanks: Governance in Comparative Perspective is the first book to explore the political and governance implications of biobanks in Europe, the United States, Asia and Australia. This book explores:

- the interrelated conditions needed for a biobank to be created and to exist
- the rise of the new bio-economy
- the redefinition of citizenship accompanying national biobank developments.

This groundbreaking book makes clear that biobanks are a phenomenon that cannot be disconnected from considerations of power, politics and the reshaping of current practices in governance. It will be a valuable read for scholars and students of genetics, bioethics, risk, public health and the sociology of health and illness.

Herbert Gottweis is Professor of Political Science and Director of the Life Science Governance Research Platform (LSG) at the University of Vienna, Austria.

Alan Petersen is Professor of Sociology, School of Political and Social Inquiry, Monash University, Melbourne, Australia. He is also an Honorary Visiting Professor at Plymouth University and at City University in London, UK.

Biobanks

Governance in comparative perspective

Edited by
Herbert Gottweis and Alan Petersen

Routledge
Taylor & Francis Group
LONDON AND NEW YORK

First published 2008
by Routledge
2 Park Square, Milton Park, Abingdon, Oxon OX14 4RN

Simultaneously published in the USA and Canada
by Routledge
270 Madison Ave, New York, NY 10016

Routledge is an imprint of the Taylor & Francis Group, an informa business

© 2008 selection and editorial matter, Herbert Gottweis and Alan Petersen; individual chapters, the contributors

Typeset in Times New Roman by
Florence Production Ltd, Stoodleigh, Devon
Printed and bound in Great Britain by
Antony Rowe Ltd, Chippenham, Wilts

All rights reserved. No part of this book may be reprinted or reproduced or utilised in any form of by any electronic, mechanical, or other means, now known or hereafter invented, including photocopying and recording, or in any information storage or retrieval system, without permission in writing from the publishers.

British Library Cataloguing in Publication Data
A catalogue for this book is available from the British Library

Library of Congress Cataloging in Publication Data
Biobanks: governance in comparative perspective/[edited by]
Herbert Gottweis & Alan Petersen.
p.; cm.
Includes bibliographical references.
1. Human genetics – Databases – Moral and ethical aspects – Cross-cultural studies. 2. Biobanks – Government policy – Cross-cultural studies. 3. Biobanks – Political aspects – Cross-cultural studies. I. Gottweis, Herbert, 1958– II. Petersen, Alan R., Ph. D.
[DNLM: 1. Databases, Genetic–legislation & jurisprudence. 2. Confidentiality – legislation & jurisprudence. 3. Databases, Genetic – ethics. 4. Public Policy. QU 33.1 B615 2008]
QH441.2.B43 2008
174′.957–dc22
2007044055

ISBN10: 0–415–42737–1 (hbk)
ISBN10: 0–415–42738–X (pbk)
ISBN10: 0–203–92799–0 (ebk)

ISBN13: 978–0–415–42737–1 (hbk)
ISBN13: 978–0–415–42738–8 (pbk)
ISBN13: 978–0–203–92799–1 (ebk)

Contents

List of Contributors

Oonagh Corrigan is Senior Lecturer in Clinical Education Research, Peninsula Medical School, Universities of Plymouth and Exeter.

Rain Eensaar is a Management Consultant and has worked for the Estonian Genome Foundation 2000–4.

Amy Fletcher is a Senior Lecturer in Political Science, School of Political Science and Communication, University of Canterbury, New Zealand.

Herbert Gottweis is Professor of Political Science at the University of Vienna where he also directs the Life Science Governance Research Platform.

Klaus Hoeyer is Assistant Professor of Health Services Research, Institute of Public Health, University of Copenhagen, Denmark.

Michaela Mayrhofer is a PhD student at the Ecole des Hautes Etudes en Sciences Sociales and the University of Vienna.

Beverley McNamara is Senior Lecturer in Anthropology and Sociology, School of Social and Cultural Studies, the University of Western Australia.

Gísli Pálsson is Professor of Anthropology, Faculty of Social Sciences, University of Iceland.

Alan Petersen is Professor of Sociology, School of Political and Social Inquiry, Monash University, Melbourne, Australia.

Barbara Prainsack is Senior Lecturer at the Centre for Biomedicine & Society at King's College London, UK.

Ingrid Schneider is a Political Scientist working as Senior Researcher and Lecturer at the University of Hamburg's Centre for Biotechnology, Society and the Environment (BIOGUM), in the Research Group on Medicine/ Neuronal Sciences.

John-Arne Skolbekken is an Associate Professor in Community Psychology at the Department of Psychology, the Norwegian University of Science and Technology (NTNU) in Trondheim, Norway.

Richard Tutton is Senior Lecturer in Sociology, Centre for the Economic and Social Aspects of Genomics (CESAGen), Lancaster University, UK.

Robert Triendl is Director of Translational Research Inc., Tokyo.

Lars Øystein Ursin is Research Fellow in Philosophy, Philosophy Department and Bioethics Research Group, the Norwegian University of Technology and Science, Norway.

Acknowledgements

We would like to thank our contributors, who responded so positively to our suggestions and who enthusiastically contributed to the authors' workshop held in Vienna in June 2006. We would also like to thank our respective universities for allowing us time and the resources that enabled us to meet and exchange ideas and to edit this volume. Grants from the Austrian Genome Project GEN-AU, and the EU's 6th Framework Programme for Research project GeneBanC have supported research for this book. The publication of the book was also supported by a grant from the Brocher Foundation (www.brocher.ch), to whom we are grateful. We would also like to acknowledge the assistance offered by a number of staff at Taylor & Francis: Grace McInnes, Kirsty Smy and Eloise Cook. Finally, we would like to thank our partners and families who have supported us in our work.

Abbreviations

AAMC	American Association of Medical Colleges
AFM	Association Française contre les Myopathies
AGRE	Autism Genetic Resources Exchange
AHEC	Australian Health Ethics Committee
ALRC	Australian Law Reform Commission
BBMRI	Pan-European Biobanking and Biomolecular Resources Research Infrastructure
BMBF	Federal Ministry for Education and Research
BRC	Biological Resource Centre
CCNE	National Consultative Ethics Committee for Health and Life Sciences
CEPH	Centre d'Étude du Polymorphisme Humaine
CSTP	Council for Science and Technology Policy
DHHS	Department of Health and Human Services
DY	Dor Yeshorim
EGC	Ethics and Governance Council
EGF	Ethics and Governance Framework
EGP	Estonian Genome Project
EGPF	Estonian Genome Project Foundation
EGRP	Epidemiology and Genetics Research Program
EPO	European Patent Office
FUGE	Functional Genomics Programme
GEN-AU	Austrian Genome Project
GPPC	Genetics and Public Policy Center
HGDP	Human Genome Diversity Project
HGP	Human Genome Project
HGRA	Human Genes Research Act
HSD	Health Sector Database
HUNT	Nord-Trøndelag Health Study
IAG	Interim Advisory Group
IASH	Israeli Academy of Sciences and Humanities
IMF	International Myeloma Foundation
IMSUT	Institute of Medical Sciences at the University of Tokyo

IRB	Institutional Review Board
JFHS	Joondalup Family Health Study
JMA	Japanese Medical Association
METI	Ministry of Economics and International Trade
MEXT	Ministry of Education, Culture, Sports, Science, and Technology
MGRP	Molecular Genealogy Research Project
NAPBC	National Action Plan on Breast Cancer
NBCC	National Breast Cancer Coalition
NCI	National Cancer Institute
NCS	National Children's Study
NEC	National Ethics Council
NGNF	National Genome Research Network
NHGRI	National Human Genome Research Institute
NHS	National Health Service
NIH	National Institutes of Health
NLGIP	National Laboratory of the Genetics of Israeli Populations
NLGIP	National Laboratory for the Genetics of Israeli Populations
NOS-S	Nordic Committee for Social Science Research
NTNU	Norwegian University of Technology and Science
OECD	Organization for Economic Cooperation and Development
OMB	Office of Management and Budget
P3G	Public Population Project in Genomics
PCR	Polymerase chain reaction
PHS	Public Health Service
PMC	Personalized Medical Coalition
PWC	Price Waterhouse Coopers
REC	Research Ethic Committee
SACGHS	Secretary's Advisory Committee on Genetics, Health, and Society
SNP	Single nucleotide polymorphism
SNTRS	Syndicat National des Travailleurs de la Recherche Scientifique
STS	Science and Technology Studies
TAB	Office for Technology Assessment
TCGA	The Cancer Genome Atlas
TMF	Telematic Platform for Medical Research Networks
USPTO	United States Patent and Trademark Office
VA	Department of Veterans Affairs
WADLS	Western Australian Data Linkage System
WAGHP	Western Australian Genome Health Project
WHO	World Health Organization

Part 1

Conceptualizing biobanks

1 Biobanks and governance

An introduction

Herbert Gottweis and Alan Petersen

In recent years there has been much discussion about the implications of new genetic developments for health and healthcare, the individual and society. According to its proponents, the so-called new genetics will benefit all of society, by reducing or eliminating disease, reducing healthcare costs, and enhancing individual choice. Research in genetics is seen to lay the groundwork for 'personalized' medicine by allowing a better match between the drug and the individual genetic profile while 'empowering' the individual by offering greater certainty about their health status and more options in healthcare decisions. In public health, a greater understanding of the contributions of genetics, lifestyle and environment to disease, derived through genetic epidemiology, is seen to provide the basis for new strategies of population-based preventive interventions. It is argued that the genetically 'susceptible' may be isolated from certain environments that predispose them to disease, or advised about changes in lifestyle that may contribute to future illness. The new genetics is surrounded by considerable hype, with reports of new genetic discoveries appearing almost daily in the news media, often accompanied by strong claims about their potential benefits for 'the public'.

Competing with these positive, utopian portrayals of new 'breakthroughs' and new therapies 'on the horizon', however, are more negative, dystopian, depictions of innovations. In the view of some writers, new genetic technologies carry substantial risks. By potentially allowing control over life itself, it is argued, genetic technologies may lead to increased surveillance and manipulation of bodies and lives, and perhaps exacerbate social inequalities and discrimination based upon biological differences. The erosion of the nature–culture dualism accompanying new genetic and other biomedical developments and its implications for concepts of self, society and citizenship has been a major theme in the recent social science literature in this area (e.g. Rabinow 1992; Petryna 2002; Rose and Novas 2005). At the same time, there has been vigorous discussion about the 'ethics' of new genetic developments and how best to regulate them in order to ensure that innovations may proceed without compromising rights and creating injustice.

In this book, we step aside from these debates and positions in order to investigate in depth a number of aspects and implications of one increasingly prominent area of new genetics developments; namely, biobanks. While biobanks are nothing new, the renewed interest in them has arisen in the wake of the mapping of the Human Genome Project and other 'gene-mapping' initiatives. The emergence of widely-publicized biobank projects internationally reflects the widespread positive expectations of new genetic developments in this context and the underlying belief in rational science and progress more generally. As we show, biobanks may take somewhat different forms and evoke different responses internationally. However, discussions and writings in this field thus far are characterized by certain prevailing concerns which mirror those in other areas of new genetics innovation; in particular a focus on the 'ethics and regulation' of developments, mostly without reference to the socio-cultural, political and historical contexts that shape these developments. Thus far, there has been little reflection on what may be learnt about the fact of their emergence, their different manifestations and operations, and various publics' responses to them for the contemporary workings of politics and power. In this book, we are concerned with exploring the governance and politics of biobanks, which we believe is fruitfully understood through reference to comparative case studies. We are aware that 'governance' has multiple, contested meanings and so one of our aims in this chapter is to clarify our understanding of this concept, why we believe it is useful and how it may be applied to this particular field.

Chapter 2 then identifies the broad shifts in governance that have accompanied and are manifest in the biobank phenomenon. However, at this point, we can say briefly that we are interested in the form in which biobanks have developed in different jurisdictions thus far; the practices which they encompass; the factors that have shaped their development; and the meanings for those who are or believe they are in one way or another affected by them. One of our aims in this book is to draw attention to aspects of biobanks that have not received sustained attention thus far and to highlight their broader implications. As the editors, we come to this project as a political scientist and a sociologist with interests in genetics and issues of governance and a belief in the value of interdisciplinary study. The contributors to this book have varying backgrounds and interests, and through their particular in-depth studies and different lines of enquiry have explored aspects of the governance of biobanks.

A number of the contributors undertook research as part of a comparative project supported by the Austrian Genome project (GEN-AU). Others contributed earlier to a special issue on 'Biobanks: challenges for "ethics"' published in *Critical Public Health*, 15 (4), 2005. All the contributors presented draft versions of their chapters at a workshop in Vienna in June 2006 and by this means we sought to allow feedback on chapters and some degree of thematic unity. Beyond suggesting a broad definition of governance we have not insisted that contributors adopt particular concepts or explore specific themes, although we have sought to include a broad range of biobank developments under way. In the event, certain recurrent themes became evident in discussions and the comparative approach revealed interesting points of convergence and contrast, which should became apparent in the chapters that follow. Before introducing these chapters and their key themes, however, we should elaborate on the biobank phenomenon and our concept of governance, and on why we have adopted the particular approach that we have.

The rise and concept of biobanks

Increasingly, biobanks have become a major strategic issue in the field of biotechnology and genomics. In the most basic definition, biobanks are collections of human biological material, often combined with personal medical, genealogical, environmental and lifestyle information, within health-care systems and the medical sciences. Biobanks come in many different forms according to the type of samples that are stored and the medical–scientific domain in which they are collected. Clinical settings, research projects and the judiciary field are typical sites for biobank collections.

Biobanks gained prominence when, some years ago, a number of countries and health providers initiated large population-based studies in order to identify genes contributing to complex diseases and to study the interface between genetic and environmental factors in the aetiology of disease. Companies involved in biobank projects, such as Iceland's deCODE Genetics, have become known worldwide to mass publics, and its CEO, Kári Steffánson, turned into an icon of the new medical genomics industry. In Britain, UK Biobank is a broadly discussed, but also controversial topic and in Estonia its biobank project has played a major role in securing this country a place in the world of science.

Currently, several European countries are establishing large biobanks with prospective collections of biological material and health data from the donors. However, the practices for collecting such materials differ widely from institution to institution, and the coordination between different biobanks is variable. There is seen to be enormous potential for health research in comparing freshly collected cell and tissue samples to old stored material, provided that the analyses give analogous results (Cambon-Thomsen 2004). In several countries, such as Great Britain, Iceland, Sweden and France, there are already large, well organized biobanks or tissue repositories representing huge populations. In addition, some countries are advantaged by having large tissue collections gathered over an extensive period of time. For example, in Norway alone, there are 15–20 million tissue samples for diagnostic purposes that go back to the 1930s. Similar patterns of collection can be seen in Germany and Austria, in particular located at pathology institutes. For most of this period there are good epidemiological registries for different diseases, especially cancer (EU 2003: 11).

Basically, two large types of biobanks are distinguished in the literature: biobanks that are based on biological specimens from patients or donors, and the population-based research biobanks that are based on biological samples from (parts of) the general population with or without disease. Population-based biobanks can be cohort studies, whereby subjects are followed over time or case-control or cross-sectional studies. The different biobanks are complementary in the sense that the population-based cohorts depend on end-points from diagnostic or disease-oriented biobanks, both for precise delineation of phenotypes and for RNA or protein analyses. On the other hand, for etiologic research questions, the researchers working with disease-oriented biobanks will need control subjects and biological material that have been collected at an earlier point in time as part of the population-oriented cohorts.

To date, biobank development worldwide has focused on biobanks based on blood samples (such as in Estonia or Iceland), whereas tissue collections have been established in a fragmented manner, resulting in tissue banks of variable size, composition, standards and with different goals (Kaiser 2002; Hirtzlin *et al.* 2003; Cambon-Thomsen 2004). However, a consensus is now emerging that the power of these resources is limited because no single resource contains sufficient samples to cope with biological or medical diversity. In recognition of the limitations of the current stand-alone biobank model, the establishment of international networks of bio(tissue)banks have been assigned a very high strategic priority not only to cover the emerging demands for such resources but also to increase efficacy in medical genomics and to reduce research costs (Pearson 2004; Hagen and Carlstedt-Duke 2004; Bouchie 2004). For example, in 2007 the pan-European Biobanking and Biomolecular Resources Research Infrastructure (BBMRI) was launched, a cooperation of all major biobanks in Europe with the goal of developing a European biology infrastructure of unprecedented scope through the cooperation of a broad variety of biobanks.

Biobank governance

In the vast majority of the literature dealing with questions of biobanks the focus is on an interconnected set of issues around the questions of informed consent, personal integrity, self-determination, confidentiality and non-discrimination. In fact, it is no exaggeration to state that these key themes of ethics and bioethics have occupied the central place in the current public and political-regulatory debates on biobanks. It seems that in public discourse – and academic literature – the main challenge in the creation and operation of biobanks is seen largely to be how to deal adequately with issues such as self-determination and confidentiality. The central question, as it has appeared in public and expert discourse, has been, how may biobanks be established and operate so as to ensure that humans continue to be protected in their rights and dignity? (Tutton and Corrigan 2004). The framing of the commodification of biobank resources as a problem of intellectual property rights is part of this tendency in the current discussion to interpret the political aspects of biobanks as a rights issue.

In this book we will take a different emphasis towards biobanks, in that we suggest a broader approach focusing on their governance. Our biobank governance perspective is comparative and we emphasize that from biobank to biobank, from region to region and from country to country this interaction displays different features, characteristics, dynamics and patterns. What makes biobank governance a complicated topic is the fact that it is not simply, as many authors seem to suggest, a matter of adopting 'the right' ethical and legal techniques and considerations that determine the fate of biobanks in society and the smooth interaction between biobanks and society. In such perspectives, society is often more or less conceptualized as 'the problem' for medical–scientific advance, and/or the adoption of ethical–legal measures understood as an 'answer to the problem'. Biobanks are identified as a novel bio-medical scientific/infrastructural development that warrants a political/legal/ethical reaction with the goal to integrate biobanks into the pre-existing fabric of regulation, medicine, law and society.

This perspective is not only narrowly technocratic but it is also misleading because it tends to ignore the complex nature of the challenge of biobank governance. As much as science is transforming society, today society has a tendency to transform science as, in particular, larger, and visible science and technology projects undergo intense public scrutiny and deliberation. There are no quick 'legal or ethical fixes' available for such changed social expectations and tendencies. At the same time, contextualizing biobanks within the given complex scientific and economic structures is as much part of biobank governance as the consideration of the question of patient autonomy. Overtly optimistic expectations with respect to the power of governance intervention are hardly met in the reality of political life. Despite manifold attempts to govern the newly emerging field of biobank development, so far the history of biobanks has been fraught with political controversy, resistance and mostly fruitless attempts to create regulatory structures for biobanks on a national level. Biobanks have turned out as rather unruly phenomena and the governance of biobanks has remained a heterogenous patchwork operating through mostly local guidelines, codes of good practice and narratives of ethics.

In this book we want to distinguish between the governance *of* biobanks and governance *through* biobanks. At the same time we maintain that the governance of biobanks is inseparably and necessarily tightly connected with governance through biobanks. Thus, we question the possibility of neatly drawing a line between the scientific–technological and the political, and emphasize the hybrid character of biobank governance. Many of the debates and legal developments dealing with biobank governance conceptualize biobanks as a challenge and topic for governance. The central issue here is the governance *of* biobanks. Typically there are two important moments when biobanks tend to be identified as problems/topics of governance in the sense of warranting policymaking, state intervention and policy actor coordination: when the financing and financial implications of biobank projects are topics of concern; and when the question of the ethical/legal/regulatory set-up of biobanks comes into consideration. The economics and the regulatory politics of biobanks are often dealt with in combination, sometimes they come into focus separately.

Cutter *et al.* have identified two basic models of biobank governance: first, legislatively created and regulated projects, such as the Icelandic Act on Biobanks establishing the framework for the Icelandic biobank, or the Estonian biobank, projects created specifically by statutes and related instruments; second, 'self-created/self-regulated' projects, like UK Biobank, which are created independently from legislation and interact with existing laws as the situations arise and that are regulated in a self-binding, but not necessarily legally binding manner (Cutter *et al.* 2004: 187–8). In both cases the issues of the establishment and operation of the biobank, the collection, handling and access to samples, and the relationships with participants, research users and society are regulated through regulations, principles and ethical guidelines, dealing with issues such as informed consent and confidentiality. Such attempts to govern biobanks develop before, while, or after biobank projects are considered, created and launched, depending on the factors such as the pre-existence or the new creation of genetic collections. In both cases the government or institutional actors close to the state operate as regulators that intend to ensure a sound interaction between biobanks and society.

One important characteristic of the governance of biobanks is the blurred boundaries between science and society that reflect the shifting relation between who is governing and who is governed. The conceptual shift in political science from 'government' to 'governance' denotes this blurring and points to the fact that today social steering is hardly anymore the prerogative of central governmental agencies. The current political arena is populated by a multitude of autonomous actors who create patterns of structured co-operation despite the absence of a central organizing authority (see Gottweis, Chapter 2). Increasingly, local and national patterns of governance blend into transnational and global forms of policymaking. At the same time, new forms of governance emerge that take on a variety of forms that co-exist in one and the same field. Unidirectional forms of governance such as 'top-down' governance or 'bottom-up' governance co-exist with multi-directional forms of governance, such as network governance. In the field of biobanks, 'traditional' governments or governmental institutions with a focus on top-down governance continue to play a crucial role in the support and regulation of research and development. At the same time bottom-up patterns of governance seem to emerge, as articulated by the increasing importance of patient groups or, more generally, by mass publics and public opinion shaping genomics related policymaking. Simultaneously, markets that steer horizontal exchanges between sellers and buyers, producers and consumers constitute important structural elements in the governance of biobanks (Rosenau 2002: 80–1). The governance of biobanks operates in this new world in which answers cannot be expected anymore exclusively from the traditional political agencies, and where governance has become multiple and contested.

But there is another important dimension of biobank governance. Biobanks are not only 'topics' and 'problems' of governance, they articulate particular rationalities and constitute a complex process of representing science, bodies, medicine and technology. They are a form of governing life and involve a multitude of actors such as scientists, patients, or industry who actively engage in building, describing and operating biobanks and who contribute to translating particular scientific–technological visions into material practices. They involve the deployment of physical infrastructures, artefacts, machines, tools, instruments and buildings. Thus, biobanks are not passive objects of governance, they constitute various narratives, representations and strategies inseparably connected with their creation and operate as structuring elements in a complex governance apparatus. The central issue here is governance *through* biobanks.

Biobanks always connect with society, culture, the economy and politics. Biobanks incorporate visions for the future of medicine and healthcare, offer resources to medical research, suggest particular interactions between medical research and the pharmaceutical industry and embed images of the patient, the citizen, collective identity and society. The precise nature of the biobank–society relationship is the result of complicated processes of negotiation, discussion, argumentation and persuasion, but also of imposition. These processes either result in funding, rules, laws and the establishment of particular structures of interaction between biobanks, society and economy, or, as we will show in this book, in the reconsideration or abandonment of particular biobank projects as incompatible with societal, economic, and/or political needs and expectations. While the governance of biobanks tends to be a process in the public realm, governance through biobanks, such as the creation of an infrastructure for personalized medicine and its implications for patient–doctor interactions operates on the level of impacts that only indirectly come into the focus of political negotiation. The ordering of this complicated relationship between biobanks, society, culture, the economy and politics is at the core of biobank governance, the topic of this book.

Emergent themes

The chapters are diverse in their foci and perspectives but they reveal a number of emergent themes pertaining to biobank governance. First, the chapters underline the high expectations and ambitions attached of such collections. Notwithstanding national differences in their approach, organization and supportive networks, the new generation of biobanks tend to be of a considerable scale, involve substantial investment from the state, industry and/or charities, and call upon the commitment and energies of a diverse array of actors. In many, if not most cases, they are viewed by decision makers as big infrastructure projects requiring huge capital investment in order to achieve the expected substantial payoffs in the future. Their proponents share the strong belief that they can deliver. While there is no certainty that biobanks will produce what is promised, they nevertheless have far-reaching transformative potential – to change the way health research is undertaken, to mobilize physical, financial and human resources, to shift relations between the individual and the state, and to reconfigure life. In their efforts to engender public support for and participation in projects, proponents have made extensive reference to the supposed future benefits that will accrue for 'the public's health' and economic development, often making strong appeals to citizen responsibilities; for example, references to 'genetic solidarity' and 'altruism' (see Petersen 2005). As the examples from Iceland, Estonia, Japan and Australia show, in some instances projects are championed and led by (and in some cases would appear to be heavily reliant on the energies of) a charismatic figure who has a strong entrepreneurial spirit and is 'media savvy'. In the information age, the effective use of public relations and diverse media is crucial if one is to 'sell the dream' – a point not lost on the new generation of biobank entrepreneurs.

The chapters also show that proponents of biobanks very often employ the rhetoric of nation-building, with expectations that developments will advance national economies and identities through participation in the global biotech economy. National pride and ambitions are often evident in biobank logos and emblems, and biobanks are often depicted in project documentation and on websites as 'flag-ship' research endeavours that are leading the way in the 'post-genomic' voyage of discovery. However, increasingly, their supportive networks transcend national boundaries. They are an example of the practice of 'big science', involving multifarious networks of researchers, policymakers, funders, and other actors (e.g. patient support groups) collaborating in the sharing of information at the international level. Earlier we mentioned the BBMRI. There is also The Public Population Project in Genomics (P3G) Consortium, whose 'Charter members' include representatives from biobank projects or cohort studies (including samples larger than 10,000) throughout the world. This has as its stated aim to promote collaboration between members of the international research community to advance knowledge transfer for the health of populations (www.p3gconsortium.org/membInfo.cfm). Like human genome 'mapping', which was reliant on international cooperation and the massive resources of the international scientific community, biobank research calls for 'cooperative competition' among nations and 'information sharing' among scientists, research institutions, policymakers and funders.

It is evident that the huge scale of biobank developments and the diverse array of stakeholder groups involved (funders, scientists, doctors, public health professionals, lay publics, etc.), combined with the prospective nature of research and uncertainties about the precise uses and users of information, poses considerable challenges for scientists and policymakers who are seeking to establish legitimacy and consent for projects. These go beyond the risks frequently associated with big infrastructure projects (airports, bridges, roads, etc.) such as the realization of ambitious plans, unforeseen engineering difficulties, cost blowouts, and so on (Flyvbjerg 2005). The fact that they deal intimately with people (as patients and donors) whose consent is needed for developments and involve diverse actor groups over a longer period, presents a problem of coordination and establishing agreement on fundamental issues. Given this, it is not surprising that the limitations and implications of established conceptions of ethics and its discourse of rights have become increasing apparent (see special issue of *Critical Public Health*, 15 (4), 2005, 'Biobanks: challenges for "ethics" '). In some countries, projects are being implemented in a context of heightened concerns about a 'decline of trust' in experts and authorities and in the regulatory systems governing biomedical and biotechnology innovations. Scientists and policymakers are acutely aware of the importance of public responses to technology innovations, of the potential for a backlash of the kind witnessed with GM crops and food, and of the need to carefully engender widespread support for projects *before* innovations are too far advanced.

The long-established conception of the science–society relationship, which is premised upon a clear separation between expert and lay knowledge and the assumption that the role of science communication should be about educating an 'ignorant public' about technology development (the so-called deficit model of public understanding) is under scrutiny (see Irwin and Michael 2003: 19–40; Wynne 2006). The question of whether new models of 'public engagement' substantially change the power relations that exist between experts and lay publics and allow an effective means for publics to deliberate on the substantive questions that biobanks gives rise to is debatable. Thus far, there has been no fundamental shift in thinking about the applicability and implications of established ethical frameworks in relation to biobanks. A continuing focus on informed consent, confidentiality, discrimination and so on, has served to deflect attention from and limit debate and policy on a range of substantive issues arising from the collection, storage and use of DNA, personal medical and genealogical information involving large samples. These include the question of whether the development of such collections represents a good use of resources and should be supported, who ultimately owns collected data and who benefits from research, and whether purported safeguards can be guaranteed. (See Gottweis' example, in Chapter 2, of the temporary changing of the law in Sweden after the Tsunami catastrophe, which allowed police the authority to match DNA from the bodies in Thailand with blood samples in the Swedish biobank.)

As is also apparent from the experiences of a number of biobank projects thus far – especially those in Iceland and Estonia – biobanks are likely to develop in unexpected ways and have unanticipated consequences. Likely confounding factors include changes in funders' priorities (resulting, for instance, from the rise or fall of the stock market or shareholder pressure, e.g. Egeen in Estonia; resistance from stakeholder groups (e.g. as with the GPs in Iceland); shifts in the interests and fortunes of biobank 'champions', inter-professional rivalries and opposition from publics and donors). As new biobanks begin to recruit and attract growing publicity, the visibility of projects will no doubt increase and there is the danger that public concern, and hence opposition, may grow. Adverse media coverage of projects or of other biotechnology developments, for example 'breakthroughs' in embryonic stem cell research, may serve to trigger public concern. During its establishment phase, UK Biobank received some adverse publicity (via the national newspaper press and via the press releases of the activist group, GeneWatch) focusing on a number of substantive issues (see Corrigan and Petersen, Chapter 9) which, if sustained, had the potential to undermine the viability of the project. As the example of the Iceland Health Sector Database reveals (see Pálsson, Chapter 3), governance is prone to failure and no amount of risk management can prevent the likely event or confluence of events, which may ultimately lead to failure.

Regardless of whether biobanks develop and deliver in ways envisaged, there is little doubt that their emergence reflects and is contributing to a general change in conceptions of health, self and society (see, e.g. Petersen 2006). In recent years, across the world new projects have developed on the premise that, in the future, medicine and healthcare *will* be delivered differently and that this *will* lead to *improvements* in health and wellbeing and that therefore biobanks necessarily operate for the broader public good. In virtually all cases, the beneficence of biobanks is taken as given. Most projects are being developed well ahead of wide-ranging debate about their purpose, their value, and their social, economic and political implications. The questions they give rise to are much broader than those typically raised by ethicists, philosophers, and lawyers and call for new perspectives and contributions from diverse disciplines and constituencies. It is with this in mind that we have focused attention on issues of biobank governance from an international comparative perspective. We hope that the chapters will help stimulate further discussion and research and prove to be a valuable source for those working in this field.

Outline of the chapters

The chapters emphasize in different ways the two dimensions of biobanks outlined above, namely as a challenge for governance and as a new form of the governance of life. Part 1 ('Conceptualising biobanks') discusses the central analytical concepts that are used in this book to develop a better understanding of biobank governance. Part 2 ('How to build a biobank: comparing different approaches') presents the main strategies that are currently deployed to set up biobanks. In this section we will analyse which factors lead to the creation of national biobank projects, and which factors encourage local (Germany) or 'bottom-up' (France) approaches and how different strategies are pursued to shape governance regimes for biobanks. As these chapters will show, governing life through biobanks takes many different forms, which points to the fact that modern biopolitics is characterized by the co-existence of multiple strategies ranging from top-down state-led approaches to commercially and patient driven approaches. Part 3 ('Biobanks, publics and citizenship') focuses on the impacts and consequences of biobanks for society, political identity, citizenship and the national body, paying cognizance to developments in particular countries. Concerns and anxieties about biobank projects have been broadly expressed in many countries. The debate about the relationship between science and society has moved centre stage. At the same time, the strong focus in the bioethical debate on informed consent, confidentiality and privacy has given way to a rethinking of the paramount position of the individual in contemporary bioethics discourse. What can be learnt about these processes from national biobanks' projects as they evolved thus far? Some biobank projects such as the ones in the UK and Australia are still under construction, other projects such as those in Israel already have a longer history of application and allow the study of the role of biobanks in a reordering of the relationship between society and medicine. The chapters will examine the population politics and forms of body surveillance associated with developments, drawing attention to some little-discussed implications of biobank developments; for example, constructions of 'the public' and the positioning of ethnic minority populations.

In Chapter 2, Herbert Gottweis sets the scene for the chapters that follow by exploring biobanks as an aspect of hybrid governance. As Gottweis explains, biobanks represent a newly emergent mode of governing life, one that entails diverse actor networks and changes in the techniques for guiding conduct. Drawing on the work of Michel Foucault, in particular his notion of biopolitics, he notes that although the state continues to play an important role in the field of biobank governance, biobanks 'constitute a new space of governance' in that they reorder relationships between different actors (patients and doctors) and institutions (industry and universities). There is a greater emphasis on 'micro-steering' and on new ways of envisioning, surveilling and monitoring the body, which has been enabled by developments in genomics and other fields, particularly information technologies. In the chapter, he outlines the tendencies which characterize governance *through* biobanks ('decorporalization', 'molecularization', 'informationization'), making reference to the practices of different biobank projects.

The following six chapters, which comprise Section 2, highlight the significant variations through which biobanks intervene in the governance of life. The importance of these contexts is clearly emphasized in the cases of Iceland and Estonia. Different national histories, cultures and political forces operate to shape developments. In some cases, as in Germany, France and the United States, contexts work *against* the development of *national* biobank projects. The chapters in this section analyze the factors leading to or impeding the creation of national biobank projects, and the factors that encourage local (Germany) or 'bottom-up' (France) approaches and how different strategies are pursued to shape governance regimes for biobanks.

In Chapter 3, Gisli Pálsson discusses the until now most publicized biobank project, the Icelandic Health Database project, which raised a range of questions pertaining to the ownership of genetic and health related biobank data and the question of patenting. In 1998, the Icelandic Parliament ratified a bill on a Health Sector Database that would assemble in digital form medical records for the Icelandic population. Later, the company deCODE Genetics, which outlined original plans for the Database, was granted an exclusive license for constructing it and using it for twelve years on the proviso that it would be returned to the Icelandic community. The project immediately became the centre of a local and international controversy focusing on ethics, privacy and social implications, in particular the commodification of medical information. The plan for the Health Sector Database has often been represented as the first of its kind, a model for others to learn from, to imitate, or to avoid. While many national and regional biobank projects that have in one way or another drawn upon the Icelandic experience are on schedule, work on the Health Sector Database itself seems to have come to a halt. Pálsson's contribution explores some of the issues that the Icelandic project has raised, in particular the one of ownership. Also, it discusses the reasons for the apparent 'collapse' of the project.

In Chapter 4, Rain Eensaar explores the Estonian biobank project that is a fascinating example for how biobanks came to be conceptualized as a form of innovation policy, as a strategy to rebuild the healthcare system, and a tool to stimulate economic growth and to build new industries. The project has experienced highs and lows during its relatively short history. The promising public–private partnership failed after three years of venture capital financing during 2001–3. Subsequently, when private investors started to postpone the payments and discuss the change of objectives, the project experienced setbacks and was discontinued until 2007 when, finally, the financing of the project was secured. But financing was not the only reason for the difficulties the Estonian project experienced. This chapter discusses other problems confronting the Estonian Genome project and in the process identifies the multiple factors that are key for the development of biobank projects.

In Chapter 5, Michaela Mayrhofer examines the biobank activities of Association Française contre les Myopathies (AFM). In recent years, the active participation of patient organizations in research activities has contributed significantly to the funding of scientific and clinical research, and enables the 'bottom up' production of knowledge on disease. Such groups would seem to represent an example of 'biosociality', or collective action based upon shared genetic identity, described by Paul Rabinow (1992). In France, as Michaela Mayrhofer shows, the AFM has played a key role in shaping French biobank development. AFM's engagement has led to the establishment of the Généthon DNA and Cell Bank, a biobank, which is entirely financed and governed by the AFM. The example demonstrates that biobank development is by no means only guided by state intervention but can also result from patients actively intervening in bio-medical development.

In Chapter 6, Ingrid Schneider asks why there is as yet no single, national biobank project in Germany. In the chapter, she examines the pre-requisites for the establishment of population-based genetic databases and identifies a number of pertinent factors working *against* a national collection. These include researchers' scepticism about any large-scale approach to biobanking and the structure of the health and insurance systems. As in the US (see Fletcher, Chapter 7) the absence of a national health system means that there is no centralized health data registry from which to access patients' data. Notwithstanding these impediments, several 'local' population-based biobanks have been established in Germany. However, to be successful, funding institutions and researchers need to portray their projects with potential public reaction in mind, since Germany has a highly-politicized culture regarding privacy and genetic research. Indeed, to be successful, biobanks have to be framed as a 'local' endeavour. Schneider explains why this is so and why narratives of national identity, utilized with other biobanks (e.g. Iceland, Estonia, UK), cannot be invoked in Germany. Whereas any nationalist framing of biobanking would attract strong criticism, not least for historic reasons, the local approach allows for de-politicization in that it obviates the need for potentially restrictive statutory measures, or demands for transparency, accountability and public participation. In the chapter, Schneider considers the implications of this politics of local biobanks for their future financing, design and research.

In Chapter 7, Amy Fletcher focuses on the initial deliberations regarding the development of governance structures for a *national* biobank project in the United States and the political and institutional factors impeding such a project. Despite the existence of a wealth of public and private biobanks across the country, the US does not have a large-scale prospective national cohort like UK Biobank. In the chapter, Fletcher describes the initiatives currently in place and attempts to tease out the relationships between public and private, and citizen and government (state and federal) that shape the development of biobank projects. As she explains, the political sphere is characterized by a multitude of contending interests and channels for citizen influence. The power of the pharmaceutical industry and the expectations surrounding 'personalized medicine' are especially salient in the US context. As Fletcher explains, this has presented particular challenges for regulators. In the chapter, Fletcher points to the obstacles to biobank governance, including the nature of the healthcare system (the lack of universal healthcare) and the absence of a related uniform record-keeping system, the federal system and the reliance on adversarial judicial processes to reconcile conflicting interests, and funding constraints. She also describes how patient groups use biobanks as resources to exert influence over research and development.

In Chapter 8, the final chapter in this section, Robert Triendl and Herbert Gottweis discuss the Japanese biobank project, 'Biobank Japan', which was initially called the 'Tailor-made Medicine Realization Project'. As in the US case, we can see in this project the significance of the driving ideal of personalized medicine. Launched officially in 2003, Biobank Japan was conceived both as a research project and as an effort that should lead, within a five-year frame, to actual outputs in the form of new therapies. Within a relatively short period of time, the biobank has become one of the largest standardized collections of blood samples linked to patient disease histories in the world. In the chapter, Triendl and Gottweis identify the reasons why the project has had a low profile and the political factors that have shaped its development. As they point out, the entire initial goal of the project, in fact, was *not* to create an infrastructure for future biomedical research but rather to develop new therapeutic approaches for a new type of medical system, based upon pharmacogenetics. Biobank Japan enjoys some of the best funding of all biobank projects compared in this volume, due in no small part to the energies and networking abilities of its leader, Yusuke Nakamura. While the low profile of the project has meant that the project has avoided many of the political controversies surrounding a number of other projects, it faces a number of issues in relation to how it utilizes the resources that have been developed. The example illustrates how the way the project was conceived and portrayed has affected some of the basic strategies and choices for collecting and analysing data and may ultimately limit the scientific impact of the project in the longer term.

The final five chapters, constituting Section 3, focus on the impacts and consequences of biobanks for society, political identity, citizenship and the national body, paying cognizance to developments in particular countries. They also highlight the significance of contextual issues on the conception and development of biobanks. Concerns and anxieties about biobank projects have been broadly expressed in many countries. The debate about the relationship between science and society has moved centre stage. At the same time, the strong focus in the bioethical debate on informed consent, confidentiality and privacy has given way to a 'rethinking' of the paramount position of the individual in contemporary bioethics discourse. What can be learnt about these processes from national biobank projects as they have evolved thus far? The chapters will examine the population politics and forms of body surveillance associated with developments, drawing attention to some little-discussed implications of biobank developments; for example, constructions of 'the public' and the positioning of ethnic minority populations.

In Chapter 9, Oonagh Corrigan and Alan Petersen examine UK Biobank and the governance implications of its approach to ethics. This project, which had a long establishment phase but finally began recruitment in 2007, is a longitudinal project designed to study the health of 500,000 of the UK population between the ages of forty–five and sixty–nine years. From its inception the public and charity-based organizers/funders recognized that such an endeavour was potentially problematic, both in terms of the public/ethical acceptability of such a project, and in securing the large numbers required for enrolment and beyond. This chapter discusses the context shaping its particular approach to ethics and governance in order to ensure consent and legitimacy for the project and to secure the ongoing participation of individuals. It discusses the significance of the 'social turn' in bioethics, oriented to managing public responses, which increasingly is recognized as a risk with such projects. This involves the greater 'engagement' of 'the public' during early phases of biobank development. Thus far, however, 'engagement' in practice has been rather limited and stakeholder-oriented and reveals the lingering influence of the 'deficit model' of the public understanding of science. Adopted strategies allow little scope for debate about substantive issues and the many uncertainties surrounding such collections.

In Chapter 10, Richard Tutton also focuses on UK Biobank. He adopts the 'co-production' framework developed within science and technology studies to examine decisions about how and in what ways ethnic minorities should be included in this project. The question of ethnic minority inclusiveness has been ignored in many studies of biobanks; however, it would seem to be crucial to the legitimacy of such collections and to their acceptance as a resource in healthcare. As Tutton explains, the issue of 'inclusiveness' impinges on matters of public trust, a point that appears to have been acknowledged by the partners of UK Biobank. However, from the outset, as is evident from the project's Ethics and Governance Framework, there has been some ambiguity surrounding the notion of 'representativeness'. UK Biobank's policies in relation to inclusiveness have been subject to criticism from various quarters, especially from those minority groups who often bear a disproportionate burden of disease. The chapter draws on data from interviews undertaken with key scientific personnel at UK Biobank (as part of a broader study of biobanks of varying size, design and purpose), exploring how 'race' and ethnicity are conceptualized, operationalized and measured. The data revealed that the scientists involved with UK Biobank were keen to stress the inclusive nature of the project, invoking notions of social justice and citizenship in the process. However, although arguing for inclusion on social grounds, the value of minority participants was seen to lie primarily in their genetic data. Tutton reports that arguments for inclusion made reference to both scientific and pragmatic factors; however, decisions involved a compromise of scientific, pragmatic and socio-political considerations. As he notes, it is too early to say whether the scientists will be successful in recruiting their target numbers of specified ethnic minority groups or that the project will be of benefit to these groups in the ways envisaged.

In Chapter 11, Lars Øystein Ursin, Klaus Hoeyer and John-Arne Skolbekken examine Scandinavian debate on the regulation of biobank research. This concerns not only the potential for direct harm to the research subjects, but also the potential indirect harm for society at large. The latter is surrounded by a remarkable ambiguity: population-based genetic research is both a source of medical hope (hence research should not be impeded), and a potential threat to existing institutions, norms and values. While sharing such concerns, the responses to these concerns have played out differently in the three Scandinavian countries, Norway, Sweden and Denmark. In this chapter the authors look at the three different attempts at regulating biobank research. They explore the politics of ethical statements drawing on white papers and biobank laws, as well as interviews with researchers and research subjects in biobank studies in each of the countries. The regulatory reasoning in the three countries share some rhetorical features: mentioning of historical instances of medical atrocities, a paternalist tradition in healthcare and scenarios highlighting future genetic discrimination. Academics have argued that genetic epidemiological research implies an intervention, which is seen as a possible threat to public health: the colonization of the life-world by medical risk discourse may contribute to the making of a community of inward hypochondriacs and outward instrumentalists. However, the victorious arguments differ between the countries, and this chapter brings out the intricate differences between legislative processes in countries usually seen as relatively similar.

In Chapter 12, Beverley McNamara and Alan Petersen focus on an Australian biobank, the Western Australian Genome Health Project (WAGHP). They examine how consent and legitimacy for this project has been framed during its pilot phase, focusing on the documents of the project, especially those of the preparatory, Joondalup Family Health Study. The project displays a number of features found in other biobank projects. As with UK Biobank (which provides a strong point of reference for the project) the WAGHP appeals strongly to notions of citizenship in arguing for public support for the project. Further, as with a number of other projects, the uniqueness of the project and its governance is emphasized. Its proponents make much of the project's consultative nature. As noted, in recent years, in the UK and elsewhere, there has been increasing emphasis on early ('upstream') engagement in relation to biomedical and biotechnology developments. In the wake of controversies surrounding GM crops and food, cloning and embryonic stem cell research, many scientists and policy makers are concerned about publics' responses to innovations such as biobanks and perceive a need to 'engage' 'the public' during the early stages of innovations. In both published documents and the community engagement strategies pertaining to the WAGHP individuals are envisaged as active participants in research. Reference is also made to a partnership and solidarity between researchers and the 'community'. However, the authors question the degree to which adopted strategies allow publics to debate the pertinent issues and influence key decisions. A strong focus on issues of privacy and informed consent in the ethical oversight of the project diverts attention from other questions relating to the project such as the ownership, use and control of the collected information. The chapter highlights a number of representational devices employed in project documentation and how these serve to restrict debate on substantive issues. The chapter concludes by drawing attention to some implications of the particular approach adopted for the project.

Finally, in Chapter 13, Barbara Prainsack examines the role of biobanks in Israel and how they serve to reinforce collective identities. In other countries that have established biobanks there have been debates about the use of ethnic categories and the proclaimed homogeneity of populations; for example, Iceland. However, as Prainsack argues, in Israel, such debates are absent, at least in the public domain. In order to understand this, one needs to appreciate the function played by ethnicity, religious affiliation and family origin in public life. In Israeli society, the maintenance of clear boundaries between population sub-groups is seen as essential for the survival of the Jewish collective. The chapter draws on data from in-depth studies of three biobanks in Israel. These involved participant observation and interviews with bio-ethicists, policymakers, biobank staff, representatives of patient organizations and 'lay' people. In the chapter, Prainsack discusses the crucial role played by science and religion in Israeli society in relation to the preservation of identity and how this is manifest in the biobanks she studied. The chapter serves further to underline the significance of social, religious, cultural and political contexts for how biobanks are conceived and operate.

References

Agamben, G. (1998) *Homo Sacer: Sovereign Power and Bare Life.* Palo Alto, CA: Stanford University Press.

Bouchie, A. (2004) 'Coming soon: a global grid for cancer research', *Nature Biotechnology*, 22: 1071–3.

Berg, K. (2001) 'DNA sampling and banking in clincial genetics and genetic research', *New Genetics and Society*, 20 (1): 59–68.

Brown, N. and Webster, A. (2004) *New Medical Technologies and Society*. Cambridge: Polity Press.

Cambon-Thomsen, A. (2004) 'The social and ethical issues of post-genomic human biobanks', *Nature Reviews Genetics*, 5: 866–73.

Crawford, R. (1984) 'A cultural account of "health": control, release, and the social body', in J. McKinlay (ed.), *Issues in the Political Economy of Health Care* (pp. 60–106). New York and London: Tavistock.

Cutter, A., Wilson, S. and Chadwick, R. (2004) 'Balancing powers: examining models of biobank governance', *Journal of International Biotechnology Law*, 1 (5): 187–92.

Dean, M. (1999) *Governmentality: Power and Rule in Modern Society*. London: Sage.

Deleuze, G. and Guattari, F. (1987) *A Thousand Plateaus*. Minneapolis, MN: University of Minnesota Press.

EU Workshop (2003) *Biobanks for Health. Optimising the Use of European Biobanks and Health Registries for Research Relevant to Public Health and for Combatting Disease. Report and Recommendations*. Oslo.

Flyvbjerg, B. (2005) *Megaprojects and Risk: An Anatomy of Ambition*. Cambridge: Cambridge University Press.

Foucault, M. (1979) *The History of Sexuality, Vol. 1: An Introduction*. London: Allen Lane.

Gottweis, H. (1998) *Governing Molecules: The Discursive Politics of Genetic Engineering in Europe and in the United States*. Cambridge, MA: MIT Press.

Hagen, H.-E. and Carlstedt-Duke, J. (2004) 'Building global networks for human diseases: genes and populations', *Nature Medicine*, 10: 665–7.

Haggerty, K. D. and Ericson, R. V. (2000) 'The surveillance assemblage', *British Journal of Sociology*, 51: 605–22.

Hirtzlin, I., Dubreuil, C., Preaubert, N., Duchier, J., Jansen, B. *et al.* (2003) 'An empirical survey on biobanking of human genetic material and data in six EU countries', *European Journal of Human Genetics*, 11: 475–88.

Irwin, A. and Michael, M. (2003) *Science, Social Theory and Public Knowledge*. Maidenhead: Open University Press.

Knoppers, B. M. and Chadwick, R. (2005) 'Human genetic research: emerging trends in ethics', *Nature Review Genetics*, 6: 75–9.

Kaiser, J. (2002) 'Population databases boom, from Iceland to the US', *Science*, 298: 1158–61.

Luscombe, N. M., Madan Babu, M., Yu, H., Snyder, M., Teichmann, S. A. *et al.* (2004) 'Genomic analysis of regulatory network dynamics reveals large topological changes', *Nature*, 431: 308–12.

Marshall, T. H. (1950) *Citzenship and Social Class: And Other Essays*. Cambridge: Cambridge University Press.

Pálsson, G. and Harðardóttir, K. E. (2002) 'For whom the cell tolls: debates about biomedicine', *Current Anthropology*, 43 (2): 271–301.

Patton, P. (1994) 'MetamorphoLogic: bodies and power in *A Thousand Plateaus*', *Journal of the British Society for Phenomenology*, 25: 157–69.

Pearson, H. (2004) 'Summit calls for clear view of deposits in all biobanks', *Nature*, 432: 426.

Petersen, A. (2005) 'Securing our genetic health: engendering trust in UK Biobank', *Sociology of Health & Illness*, 27 (2): 271–92.

Petersen, A. (2006) 'The genetic conception of health: is it as radical as claimed?', *Health: An Interdsiciplinary Journal for the Social Study of Health, Illness and Medicine*, 10 (4) (10th Anniversary Issue on 'conceptions of health'): 481–500.

Petryna, A. (2002) *Biological Citizenship: Science and the Politics of Health after Chernobyl*. Princeton, NJ: Princeton University Press.

Plummer, K. (2003) *Intimate Citizenship: Private Decisions and Public Dialogues*. Seattle, WA and London: University of Washington Press.

Public Population Project in Genomics (P3G) Consortium. Available at: www.p3gconsortium.org/membinfo.cfm, accessed 21 August 2007.

Rabinow, P. (1992) 'Artificiality and enlightenment: from sociobiology to biosociality', in J. Crary and S. Kwinter (eds), *Incorporations* (pp. 234–52). New York: Urzone.

Rabinow, P. (1999) *French DNA: Trouble in Purgatory*. Chicago, IL: University of Chicago Press.

Rosenau, J. N. (2002) 'Governance in a New Global Order', in D. Held and A. McGrew (eds), *Governing Globalization: Power, Authority and Global Governance* (pp. 70–86). Cambridge: Polity Press.

Rose, N. (2001) 'The politics of life itself', *Theory, Culture & Society*, 18: 1–30.

Rose, N. and Novas, C. (2005) 'Biological citizenship', in A. Ong and S. J. Collier (eds), *Global Assemblages: Technology, Politics, and Ethics as Anthropological Problems* (pp. 439–63). London: Blackwell.

Rothstein, M. A. and Epps, P. G. (2001) 'Ethical and legal implications of pharmacogenomics', *Nature Review Genetics*, 2: 228–1.

Tutton, R. and Corrigan, O. (2004) *Genetic Databases: Socio-ethical Issues in the Collection and use of DNA*. London and New York: Routledge.

Waldby, C. (2000) *The Visible Humam Project: Informatic Bodies and Posthuman Medicine*. London and New York: Routledge.

Waldby, C. (2002) 'Stem cells, tissue cultures and the production of biovalue', *Health: An Interdsiciplinary Journal for the Social Study of Health, Illness and Medicine*, 6 (3): 305–23.

Webster, A., Martin, P., Lewis, G. and Smart, A. (2004) 'Integrating pharmacogenetics into society: in search of a model', *Nature Reviews Genetics*, 5: 663–9.

Wynne, B. (2006) 'Public engagement as a means of restoring public trust in science: hitting the notes, but missing the music?', *Community Genetics*, 9: 211–20.

2 Biobanks in action

New strategies in the governance of life

Herbert Gottweis

Introduction

Starting with the great public and scholarly interest in the Icelandic Health Sector Database, and then with other genetic database projects such as those in Estonia and in the United Kingdom, biobanks have become a much debated topic over the last decade. This discussion raises the question of what it is biobanks are 'doing' that gets them so much attention? In this chapter I will argue that one reason for the strong interest in biobanks is a growing understanding that biobanks constitute a new strategy in the governance of life. Biobanks question and transform the boundaries between the scientific/technological, the social, the cultural, and the political, and thereby can be interpreted as moments in a specific restructuring going on in the domain of life. It is in this respect that we can talk about a newly emerging governance of life *through* biobanks. I will argue that biobanks are not only an object or topic of governance, such as in regulation policies, but they can also be seen as something through which the governance of life operates. Their representations of and interventions in life constitute a new, heterogeneous space of governance in which, for example, relationships between patients and doctors, between genes and disease, scientists and the public, the pharmaceutical industry and medical sciences, or images of collective identity are defined and re-defined.

Biobanks are by no means new in the world of medicine and biological research. The systematic collection of human cells and tissues has been going on for many years, even dating back to the nineteenth century, including fixed and processed as well as frozen viable and non-viable material. In Europe and in many countries, millions of tissue samples are being permanently stored, for example, in the context of pathology institutes. Only relatively recently were large patient registries and population surveys initiated, enabling the coupling of biological and genetic data, and general patient data. But these 'early' repositories had very different operational and scientific goals to the current biobank projects. Pathology collections, like today's biobanks, were oriented to identifying and understanding diseases. However, medical doctors in the past, for example, in the nineteenth-century Hapsburg monarchy working at the University of Graz, which held a central pathology collection of the Empire, approached questions of disease differently than Swedish medical doctors at Umea Hospital today. Needless to say, then, the phenomenon of biobanks must be located within a larger transformation in the biological sciences. Two major areas of life science development have led to a new impetus to create biobanks or use old biobanks in new ways: one, the methodological breakthroughs in molecular biology and proteomics with possibilities to handle large databases, and, two, the systematic collection of fresh material from large populations. These transformations and their application potentially began to raise completely new sets of issues. Furthermore, biobanks, located at the disciplinary intersection of medical science, genomics, genetics, molecular biology, and informatics can also be well contextualized within a new discourse and knowledge on the body and health. They promise a new and systematic approach towards disease and drug development based on insights from genomics, proteomics, and pharmacogenomics. Among other things, they claim to predict the likelihood that an individual will develop a disease so that pharmaceutical drugs could be used to prevent its onset rather than resorting to treating the symptoms once a disease has developed. Lifestyle advice could be targeted to those deemed 'genetically susceptible'. Based on evidence from biobank research, pharmacogenetics could help to improve the efficacy and safety of medicine. Thus, biobank policies would constitute a major effort in establishing a preventive and, so the story goes, much more cost-efficient approach towards medicine.

Thus, the emerging landscape of biobanks is hardly a phenomenon of only local interest, and a clear realization in the life-science community exists that the creation of worldwide biobanks networks and cooperation will constitute a crucial step in rebuilding the genomics/postgenomics apparatus of modern biotechnology. The policy vision behind this development is that the exploitation of biobanks and registries in Europe and elsewhere is crucial during a period when recent improvements of large-scale research in cell and molecular biology will enable new possibilities for health research, knowledge production, and understanding of causes, progression, prognosis, and treatment of different diseases (Berg 2001). Ultimately, so the narrative told in many locations goes, biobanks might be an important step towards the improvement and development of preventive, genetic, and 'personalized' medicine. In fact, in some countries, such as Japan, biobank projects are seen as 'implementation' of the idea of 'personalized medicine', understood as the development of new, 'tailored' drugs based on the study of diseases and drug side effects, made possible by genetic database research (see Triendl and Gottweis, Chapter 8). Such representations of the operation and impact of biobanks must be seen as important contributions in structuring the field of action for biobanks and, thus in the ways biobanks govern life.

Consequently, what biobanks are 'doing' goes far beyond contributing to basic research in biology. They are increasingly regarded as large, biological infrastructures with a broad field of application and connected to a variety of scientific, economic, and political objectives. At the same time, these objectives and the precise strategies to reach them often remain relatively vague. To some extent, biobanks act like a 'machine to make a future', to borrow the concept from Francois Jacob and Paul Rabinow, large mechanisms or infrastructures designed to give unknown answers to questions the experimenters themselves are not able to clearly ask (Rabinow and Dan-Cohen 2005: 4). In this chapter I look closer at these 'future machines' and discuss some associated new methods in the governance of life.

Biobanks and the state

Today's biobank projects are not mainly territorialized and ordered along the idea of the modern nation-state or characterized by the dominant role of an all-powerful state. There can be no doubt that in the United Kingdom, Iceland, and Estonia the state is an important actor in the field of biobank governance. Also, these projects clearly have strong national dimensions and see themselves in the service of improving the health of their populations. However, we can also recognize a number of novel political features in current biobank development that need to be emphasized. Biobanks, 'banks of life', can be seen as constituting a way of organizing life, of collecting, storing, interpreting, and assembling life in the form of human materials, such as tissue or DNA. They articulate a particular form of biopolitics, a politics in which the 'bios' described by Aristotle, as the qualified life of a legally protected citizen and member of a community, connects in a novel, particular way with politics.

Reflection on the special relationship between life and politics is summarized famously in the term *biopolitics* and is closely related to the seminal work of Michel Foucault. Foucault has pointed to the significant historical transition contemporaneous with the shaping of industrial capitalism, in which emphasis shifted from the primacy of sovereignty, law, and coercion or force 'to take life' to the development of new forms of power constitutive of life. Such processes of subjectification can occur in the form of the subjection of individuals to techniques of domination or through subtler techniques of the self. This power of life co-evolved in two forms: disciplining the body and regulating populations. Whereas the former had as its object the individual, the latter addressed itself explicitly to the 'ensemble of the population' as a field of shaping and forging. These two strategies constituted the two poles around which the power over life was organized. The then-emerging biopolitics focused on the administration of life, in particular, on the level of populations and was concerned with matters of life and death, with birth, health, illness, and other processes sustaining the optimization of the life of a population (Foucault 1979; Dean 1999: 99). The activities of government and the state involved collecting, collating, and calculating data on the characteristics of the population (births, deaths, rates of disease, etc.) to be complemented by those on individuals who engage in practices of 'self-government' (Rose 2001).

Despite the fact that bio-power was not an exclusive project of the state, but worked through a variety of institutions such as family, hospital, and the human sciences, Foucault argued, the state nevertheless played a central role in coordinating and steering biopolitics. At the same time, Foucault interpreted the locus of intervention of biopolitics to be the human body, conceived as a more or less coherent, whole entity. Related to the idea of disciplining and steering human bodies was the idea of panopticism as the essence of social control (Foucault 1977), that is, the desire to direct behaviour through the imposition of a totalizing and instrumental rationalism (Sewell 1998). This desire incorporated a need to know as much about individuals as possible, a need pursued through the deployment of instruments of measurement, enumeration, and rationalization. Such intense scrutiny not only extracted information about the activities of individuals, but it also went a long way towards shaping their subjectivity as individuals who came to see themselves in the ways they are defined through surveillance (Sewell 1998). Not only were the bodies of modern biopolitics 'coherent wholes' and constituted and controlled through methods of disciplining and surveillance, or guided through care of the self, these bodies were also territorialized in the context of the modern nation-state.

This government of life operated within the space constituted by the state and was defined through the idea of the nation, in whose name and in defence of its population modern wars were waged. This modern form of biopolitics, as Giorgio Agamben (1998) convincingly has argued, was not just characterized by a 'productive' attitude towards life, the shift from 'taking away life' to 'generating and preserving life', be it through the shaping of modern health policy or the modern state's role in the 'rise of the clinic'. To Agamben, the 'state of exception', the sovereign decision of excluding people from the realm of the law by stripping them of their rights, remains a constitutive feature of contemporary state power. 'Bare life', the reduction of certain articulations of life such as lacking individual and political rights, has the most intimate link to sovereignty. The exception from the rule of law, through its suspension, is what ultimately grants the rule its legitimacy (Agamben 1998).

These features of what we could call 'old' biopolitics seem to be questioned, or at least in need of thorough reconceptualization if we think of a number of contemporary trends in today's 'new' government of life. The recent surge in biobank projects and initiatives represent illustrative examples of this development, as they seem to indicate a new phase in the governance of life. Body surveillance in the context of the developments of contemporary life sciences means something distinctly different than in earlier times, with respect to *surveillance*, with respect to *bodies*, and with respect to the shaping of infrastructure of surveillance and monitoring – the shaping of the structure and organization of institutions of monitoring bodies and populations.

Biobanks actively contribute to this process of transformation. They are, in a way, laboratories where these processes can be observed, and thus become an integral element of what may be termed life *governance*, and thus the shaping of new spaces of governance. But, as I will show, the new strategies in the government of life remain embedded in a strategic repertoire in which 'classical' biopolitical strategies continue to be of relevance.

Decorporalization, molecularization, informationization

To start with, biobanks are an important expression in the tendency towards *decorporalization* in modern biopolitics. Biobanks represent a new politics of disappearing bodies. The body of biobanks is an inherently decomposed body (Brown and Webster 2004), a body split into systems and collections of blood, proteins, serums, genes, and SNPs. The elements of the decomposed body in a biobank obtain their relevance through their connection with each other in the system of the biobank. What we observe here is a management of living fluids and living cells that do not represent other, larger bodies but form their own bodies. Thus, biobanks create a new 'bodily' phenomenon and a new structure for moving bodies and their parts and establishing relationships between them. In the Japan biobank project, the central goals are to assemble blood samples and DNA as well as clinical information about 300,000 individuals, based on standard 'informed consent' protocols to store this information in line with appropriate data safety measures; to determine specific groups of symptoms or reactions to medication using the clinical database and to perform a SNPs analysis covering all genes; and to develop appropriate software tools for the analysis, and application of the various datasets created by the project. The goal of the project, therefore, is to create a database that can be used to determine the genetic basis of drug susceptibility and to identify genetic traits related to disease susceptibility, disease progression, or responsiveness to certain forms of therapy. 'Surveillance commences with the creation of a space of comparison and the introduction of breaks in the flows that emanate from, or circulate within, the human body' (Haggerty and Ericson 2000: 612). The key issue is not to monitor 'complete' bodies, but collections of blood and DNA that can be associated with subgroups in the population. The new biopolitics increasingly is a politics of the dispersed body, to the extent that we lose entirely sight of the 'complete body' (Mayrhofer and Prainsack 2007).

An important dimension and precondition of decorporalization are *molecularization* and *informationization.* Molecular biological approaches, advances in computer and information sciences, and the convergence of these two domains have also led to a fundamental reconceptualization of health and disease in medical discourse. Contemporary biology is characterized by its strategy to study biological phenomenon at the level of macromolecules. Since the 1930s, the focus of modern biology has begun to shift from the cellular to the subcellular level of biological phenomena. In the practice of modern biology the locus of life phenomena was conceptualized at the submicroscopic level (Kay 1993). Subsequently, with the shift from the protein paradigm to the DNA paradigm during the late 1950s, explaining the operation of DNA and understanding and handling its manipulation moved to the centre of biological and medical interest. During the 1970s, genetic engineering, then, during the 1980s, genetic testing, gene therapy, and the human genome project seemed to delineate a new future for biomedicine whereby knowledge and mastery of DNA were the keys to solve many of the important problems in medical research (Nelkin and Lindee 1995).

Informationization refers to the process by which information technologies have transformed practices and organizational features of fields such as medical research. The rise of biobanks as a key tool of medical research and drug development is inseparable from computerization and highly sophisticated information technology. In 1998, the Icelandic Ministry of Health announced its plans for the construction of a Health Sector Database on the entire Icelandic population. These plans, initiated by the private company deCODE Genetics, specified how and under what conditions to assemble medical records – and, possibly, combine them with genetic data and genealogical records for the purposes of tracking the presumed genetic bases of diseases and economizing the National Health Service. The founding premise of deCODE Genetics was that although the nuclear family has proved to be a useful unit in the study of 'monogenic' disorders, for complex or 'polygenic' disorders that are 'sporadic', skipping generations, more information and higher resolutions are needed. Researchers from deCODE reasoned that a Health Sector Database might be useful in this context. Although information on DNA, medical data, and genealogical records would only be combined in the context of specific research projects and would be monitored by ethics committees and public officials, their synergistic coexistence was supposed to enhance each other's economic and medical value (see Pálsson, Chapter 3). These developments were inseparable from the rise of bioinformatics, multidisciplinary research at the interface between informatics and biology, with additional input from statistics and mathematics. In parallel to bioinformatics, several related and partially overlapping research areas have evolved, such as computational biology mathematical biology and biostatistics that all are key to biobank research. Modern genetic databases are sites where molecularization and informationization crystallize in the form of biological infrastructures that create new trajectories of monitoring bodies.

Bodies monitored and commodified

As previously discussed, biobanks are located at the disciplinary intersection of medical genomics, genetics, molecular biology, and informatics, and they belong within a new discourse on the body and health, within which new trajectories of body monitoring materialize. Because the new discourse on body and health promises a new and systematic approach towards disease and drug development, lifestyle advice and pharmacology could be targeted to those 'genetically susceptible' to prevent the onset of certain diseases, which could in effect reduce costs. Biobanks, then, actually or potentially could contribute to the contemporary rebuilding of practices in healthcare, the categorization of patients, the definition of disease and susceptibility, the determination of points of entry for treatment, and the emergence of new actors in the field of health policy.

Together with decorporalization and molecularization comes a tendency in political rule towards the transition from *macro-steering* (such as by the state) to *micro-steering,* the rise of new actors in medical research and healthcare, and the increased importance of self-steering. Biobank policies and initiatives must be understood in close relationship to a fundamental change in current health policy practices and disciplinary transformations in medical research. The micro-steering of biobanks brings a variety of new actors into the governance structure that have a substantial impact on the financing and operation of biobanks, but also of opportunities of access and of the definition of social centres and peripheries related to biobanks. While some social actors such as private investors or patient collectives gain privileged access to biobanks, others remain excluded.

During the 1980s and 1990s throughout the Organization for Economic Cooperation and Development (OECD), privatization had been introduced as a tool to increase both allocated and technical efficacy in the financing and provision of medical services and to meet patient preferences for individual choice in healthcare. Unlike in earlier years, national healthcare systems no longer prefer solutions that have grown out of their historical, institutional, and cultural environments, but instead adopt similar approaches cross-nationally. The result of this has been the increasing integration of management-concepts in the healthcare sector throughout the OECD (Oxley and MacFarlan 1995; Stewart 1999). Healthcare reform, which focused on macro-managing for a long time, turned to micro-managing. Costs, insurance rates, efficiency, effectiveness, and incentives have become important concepts. Corporatization and commodification of healthcare and medicine are the result of the moves by private corporations to appropriate increasing areas of the healthcare sector under private ownership and/or management.

Simultaneously, private companies, often financed by venture capital, have become key actors in the field of genomic medicine driving the dynamics of the field. The gradual establishment of a regime of appropriation via patenting has been a central technique for defining the structure of the emerging medical-genomics apparatus. Ever since the landmark 1980 US Supreme Court decision in *Diamond v. Chakrabarty*, which ruled that genetically modified bacteria were patentable, intellectual property rights have become a key topic in genomic medicine. Today the US Patent and Trademark Office (USPTO) holds that even 'natural' DNA as a chemical compound once isolated and purified is patentable. Thus, the PTO and the European Patent Office (EPO) have treated isolated and purified nucleoide sequences as if they were the same as bio-chemicals (Andrews 2002: 803). To date, the PTO has patented approximately 20 per cent of the human genome, or about 4,500 genes (Jensen and Murray 2005). Human Genome Sciences has filed applications encompassing 7,500 full-length genes. The French company Genset SA has generated over 90,000 sequences of 5-prime untranslated region sequence tags of human full-length cDNA clones, and patent applications will be filed. Today we observe in health and medical policy a tendency for the state to pull out of financing and decision making, and new actors, ranging from healthcare providers, patient groups, citizen groups, and private companies to move into the centre of health and medical policy decision making. This social reorganization of the healthcare sector has also shaped the dynamics of biobank strategies.

In Iceland, the company that led the shaping of the country's Health Sector Database was deCODE, physically located in Iceland but registered in the United States. In Estonia, the Estonian Genome Project was initially funded by the private company EGeen. In the United States, private healthcare providers such as Marshfield Clinic or companies such as Genomics Collaborative Inc. play a central role in organizing biobank projects. In France, the private non-profit sector patient organization, Association Française contre les Myopathies (AFM) is identifiable as *the* major actor in the field of biobanking. (See Pálsson, Chapter 3; Gottweis, Chapter 2; Eensaar, Chapter 4; Fletcher, Chapter 7; Mayrhofer, Chapter 5.) In total, the AFM runs fourteen biobanks and collections around the world and is extremely active in the areas of genetic research, patient care, and legal issues.

At the same time, biobanks have also been conceptualized as a mechanism to promote international competitiveness. Among other things, they are seen as being capable of significantly influencing knowledge industries and creating the competitive advantage of certain regions or countries. It is often argued that Europe's national healthcare systems seem to have a strong advantage in particular vis-à-vis the United States, where the absence of a national healthcare system is seen as an obstacle for population-based studies complemented by health data. In Estonia, for example, the Estonian Genome Project has been presented as a potential catalyst for the national biotechnology industry. While governments continue worldwide to be major actors in biobank initiatives, private and non-governmental actors have come to assume a crucial role. At the same time, in some countries such as Israel and Iceland, narratives of genes as national assets co-exist with privatizing tendencies in biobank development.

As a result of these and other developments, biobanks are closely associated with the rise of a new *politics of biovalue*. Catherine Waldby has defined biovalue as 'the surplus of *in vitro* vitality produced by the biotechnical reformulation of living processes' (Waldby 2000; 2002). Tissues can be leveraged biotechnically so that they become more prolific or useful, through processes such as the fractioning of blood, the use of polymerase chain reaction (PCR) for the amplification of genetic sequences, the creation of cell lines, genetic engineering, or cell nuclear transfer. The biovaluable engineering is often associated with the requirements for patenting, so that surplus in vitro vitality may eventually be transformed into surplus commercial profits, as well as in vivo therapies (Waldby and Mitchel 2005). Needless to say, the issues of ownership and patenting have become major topics in the discussion on genetic databases. Access, control, and ownership of biobanks and their applications are in particular at the centre of those projects where private industry plays an important role.

In Iceland, deCODE Genetics received the license to construct the genetic database in return for a fee paid to the medical service. A rival biotechnological company, UVS, was established in the heat of the debate on the Health Sector Database, partly to challenge deCODE's 'monopoly' of biomedical research in Iceland. Much of the discussion of the database project focused on issues of property, ownership, and control. The arguments of the proponents of the Icelandic project tended to emphasize the opportunities it provided in terms of medical advances, work, entrepreneurship, and private initiative, in the age of challenging 'new economy' and stagnant or declining fishing stocks. A fundamental debate developed concerning the ownership of and access to genetic information and medical records. The property issue has often been discussed with reference or in comparison to ongoing debates about another thorny common-property issue, namely, the allocation of individual transferable quotas to rights in fish. For many of the critics, the commodification of biomedical data was problematic. The dominant focus in the debate was the fact that a private multinational company proposes to explore the genetic bases of common diseases in the entire Icelandic population and to commercialize its results. In Estonia, the preparation and establishment of the Gene Bank was funded by investors through the limited company EGeen, which was granted an exclusive commercial license. But the Estonian Genome project was construed as more than just a large research project; it was defined as a way of pushing Estonia's post-Soviet economy towards Western standards (see Eesaar, Chapter 4; Gottweis, Chapter 2).

Thus, commodification, patenting, the rise of new, private actors in genomic medicine, and the new politics of biovalue are integral parts of the new politics of biobanks. An important new feature in the governing of life through biobanks is that today's biobanks are not simply machines easily taken into service by the state. Biobanks are not only engaged in a process of monitoring *bodies*, or at least dispersed bodies, but they themselves also constitute *monitoring* bodies. However, just as these monitoring bodies deal with fractionalized, decomposed bodies, they themselves display a *rhizomic* character. It seems that modern biopolitics is not dominated by a strong state, but by highly decentralized, but nevertheless interrelated, *rhizomic assemblages* (Deleuze and Guattari 1987). Such assemblages consist of a multiplicity of heterogeneous objects whose unity comes from the fact that these items function together, that they work together as a functional entity. They comprise discrete flows of people, signs, and chemicals, knowledges, tissues, genes, and institutions (Patton 1994: 158; Haggerty and Ericson 2000: 607). The focus of such networks is not the disciplining or control of bodies but the transformation of the body into information and binary codes so that they can be rendered more mobile and comparable (Haggerty and Ericson 2000: 613). Biobanks are structured in precisely this manner: the trend is to form large gene collections in order to explore comparatively small bits, which, it is hoped, will grant insight into the genetic basis and the contribution of gene-environment interactions to disease (Cambon-Thomsen 2004: 867). Ideally, the combined insights will result in predictive models that relate genetic disposition to drug development and drug consumption. As the Japanese biobank project demonstrates, biobanks are not necessarily located in one site, but across several sites; they comprise a broad array of actors, artifacts, and machines that are assembled and reassembled in the project. The result is a highly complex infrastructure that spans hundreds of local hospitals to the University of Tokyo and the Reiken Center. From there, in the form of the HapMap project (see below), this infrastructure of collection interrelates with data collections in other countries and assumes transnational character.

Thus, the 'new' government of life in biobanks also seems to be characterized by its deterritorialized, *transnational/global* character, which neither excludes national projects nor a national rhetoric. But neither the nation-state nor populations are its main or exclusive technical-scientific point of reference, but rather global/transnational networks and assemblages. In general, increasingly more technological and scientific advancements in medical genomics are the result of collaboration on a global scale. For example, although Celera Genomics completed the sequencing of the human genome, and the US government sponsored HGP and the core of the work was performed in the US and the UK, the international human genome sequencing consortium included scientists at sixteen institutions across France, Germany, Japan, and China. A similar picture develops in the biggest follow-up project to the sequencing effort of the 1990s: the HapMap project, which has significant ties to a number of biobank projects, such as the Biobank Japan project. Launched in 2002, it charted genetic variation within the human genome. The central idea of the HapMap project was to create a human haplotype map. The HapMap project's goal was to allow geneticists to scan the entire genome rapidly for disease genes by analysing some 300,000 SNPs (as compared to the some ten million common SNPs scattered throughout the human genome). To create the HapMap, DNA will be taken from blood samples from Nigeria, Japan, China, and the United States with public funding from agencies scattered all over the globe, from the Japanese Ministry of Education, Culture, Sports, Science and Technology to Genome Canada in Ottawa, the Chinese Academy of Sciences in Beijing to the US National Institutes of Health in Bethesda. In the field of biobanks, the structure is currently still dominated by a national branding, such as 'Biobank Japan' or ‘UK Biobank’; however, other projects such as the Icelandic and the Estonian biobank project already display strong transnational features in the financing structure and product development strategy. At the same time, there is much discussion and preparation going on at the transnational and EU level to connect the various biobank projects, the leading protagonists, infrastructures, and strategies in a common effort, such as the emerging Biobanking and Biomolecular Resources Research Infrastructure (BBMRI) initiative. Thus, it seems that much activity in the new government of life has expanded into the global arena where new forms of linking local and global fields of operation seem to be developed.

Finally, biobanks and their expressed goal to contribute to the development of a ‘personalized medicine’ can be seen as part of a larger tendency in the medical/health system to shift the burden of health responsibility from macro-actors such as the state to the individual level. Today, health is increasingly discussed throughout the West in terms of self-control and framed within the language of an ethics of health that requires the disciplining of the individual conduct of life (Crawford 1984: 72–6). The creation of cohorts of susceptibility and risk in the discourse of personalized medicine is part of this strategy. This managing of the self (Foucault 1979) is also reflected in a multitude of technical and organizational novelties within healthcare, where managed care is the most important and most paradigmatic example. In the past, health policies were based on the collection and tabulation of numerical information about populations and provided the rationale for hygienic strategies. Likewise, strategies to minimize risks of the environment, labour conditions, or the maintenance of the body were central elements for public health strategies. Although these strategies continue to be important, the focus of strategies has begun to shift from the group to the individual level. As has been argued by Nikolas Rose, the ideal of the omnipresent state that would shape, coordinate, and direct the affairs in all sectors of society has lost its grip on the public imagination. Accordingly, in the health field concentration has moved from ‘society as a whole’ to ‘risky individuals’, individual susceptibility (to genetic disease, for example), and, accordingly, to ‘risk groups’ (Rose 2001). The proactive management of the human body has become a core element of collective and individual strategies of health maintenance. Personalized medicine, with biobanks as one of its key instruments to create cohorts of susceptibility as points of reference for individual orientation, is part of this tendency in the new biopolitics towards self-steering. It was no coincidence, for example, that the ‘Biobank Japan’ was first called, the ‘Tailor-made Medicine Realization Project’. Launched in 2003, Biobank Japan was conceived both as a research project and as an effort that should lead, within a five-year frame, to actual output in the form of new therapies. In fact, the entire initial goal of the project was *not* with creating an infrastructure for future biomedical research but with the development of new therapeutic approaches for a new type of medicine, ‘personalized’ medicine.

Biobanks, collective identity, and the future of the human body

These new features of contemporary biobank development – decorporalization, molecularization and informationization, micro-steering, the politics of biovalue, their rhizomic character, transnational/global orientation, and a new politics of self-management arsing as a related healthcare paradigm – allow us to see better what it is biobanks are 'doing' today, and to characterize them as a new strategy in the government of life. A further important aspect of biobanks is their contribution to changing the definition of citizenship and collective identity, thereby constructing and positioning individuals and their legitimate rights and claims in the political space.

As several chapters of this book show, biobanks play a key role in delineating collective identities. Biobanks are to some extent state projects, but not exclusively so. New organizations, communities, patient groups, self-help organizations and a multitude of newly emerged actors in the field of private insurance seem to have occupied spaces hitherto filled by the state. In France, for example, the AFM is a good example for patient groups actively intervening in biomedical research, among other means by organizing a biobank. In Iceland and Estonia, private corporations take care of a fundamental reorganization of the healthcare system.

Thus, biobanks and how they are explained and justified seem to be part of the emergence of what Adrina Petryna has called 'biological citizenship' (Petryna 2002), which underlines the increasing significance of specific biological presuppositions in conceptions of what it means to be a citizen (Rose and Novas 2005: 440). Building on Marshall's work on citizenship (Marshall 1950), Rose and Novas argue that, increasingly, biological images, explanations, values, and judgements get entangled with a more general contemporary 'regime of the self' as the prudent individual who actively shapes his or her life through acts of choice. Thus, citizenship can be understood as an 'evolutionary process' with civil rights granted in the eighteenth century, to be continued with the extension of political citizenship in the nineteenth century and of social citizenship in the twentieth century. In its collective expression, biological citizenship is articulated in new forms of 'biosociality', collectivities defined by categories of corporeal vulnerability, genetic risk and susceptibility (Rose and Novas 2005: 441–2).

In bioethical discourse, the issues of informed consent, personal integrity, self-determination, confidentiality, and nondiscrimination play a major role in current debates about biobanks and convey the image of individual citizens actively taking care of their rights and needs in the context of biobank development and practice. An example of this is the case of breast cancer testing, a field of central importance for tissue biobanks. The identification of the BRCA1 gene in late 1994 created a fascinating political dynamic around the question of how to define the challenge of BRCA1 for regulatory policy making. In the United States, many powerful advocacy groups, including the National Breast Cancer Coalition (NBCC), got involved in the discussions. The Hereditary Susceptibility Working Group of the National Action Plan on Breast Cancer (NAPBC), a public–private organization, brought together activists and scientists with funding from the NIH. An NAPBC Working Group recommended the example of the Colorado state legislature for the national level to ensure that genetic information was the private property of the individual alone. It recommended that insurers be prohibited from access to genetic information, while all holders of genetic information should be prohibited from releasing genetic information. In particular, the NAPBC emphasized the unique nature of genetic information (Parthasarathy 2007). Thus, the struggle was not directed *against* the identification of a particular group as defined by a particular gene, but about rights of a particular, biosocially defined group. But the 'new biopolitics' as we observe it emerging in the field of biobanks is not only about biological citizenship, the rise of biosocial actors and moments of self-guidance, and social reinvention through biology.

One difficulty with Marshall's concept of citizenship, however, is its 'evolutionary tendency', the idea of the unfolding of some sort of specific order following a specific logic. Thus, it is crucial to understand citizenship not as an essential concept but as a concept describing a loose conglomerate of spheres of action in which communities are developed that attribute certain 'contestable' rights and responsibilities to human beings (Plummer 2003: 56). The rise of biological citizenship surely can be understood as an important aspect in new government of life through biobanks. But active (biosocial) citizens who insist on their interaction with the biomedical system based on the respect of their rights as citizens and human beings is only one facet of the currently emerging biopolitical order.

To be sure, current bioethical and legal discourse in the field of biobanking literally conjures images of the human being of modernity, the coherent self-determined and rational individual equipped with human and individual rights. At the same time, the applied medical–scientific practices and technologies seem to deeply question and undermine this eighteenth and nineteenth century version of the human subject. The 'politics of the dispersed bodies' can also be seen to be in conflict with the conjuring up of an image of the person that might have been lost in the data storage systems of contemporary biobanks. Key medical ethics principles such as confidentiality or informed consent face huge difficulties in the implementation of large biobank projects with shifting purposes and goals that were unknown when they were initially launched. In a way, the disappearance of bodies in biobank systems has raised the question of the disappearance of the modern citizen or patient as we know him and her.

This situation has led to different reactions. The obsession with informed consent (Brekke and Sirnes 2006), confidentiality, and privacy in bioethical and legal discourse in the context of biobank regulation was one important reaction. But recently, the bioethics community has also begun to question the idea of the modern human being as its benchmark. In a remarkable paper published in *Nature Review Genetics* by two of today's leading bioethicists, Bartha Maria Knoppers and Ruth Chadwick, we can already read that today there are more and more calls 'to rethink the paramount position of the individual in ethics'. The authors continue in their discussion by approvingly quoting a recent World Health Organization (WHO) report on genetic databases that states: 'The justification for a database is more likely to be grounded in common values, and less on individual gain . . . It leads to the question whether the individual can remain of paramount importance in this context' (Knoppers and Chadwick 2005: 75). Knoppers and Chadwick then go on in the development of their argument by discussing 'new ethical principles' such as reciprocity, mutuality, and solidarity as possible strategies to go beyond the more traditional 'individual-centered' approaches in ethics. It seems that in the face of mounting difficulties in biobank projects to deal with individual-centered approaches in ethics, such as in informed consent, bioethical ideology has already begun to develop a 'new pragmatism' (Knoppers and Chadwick 2005: 78).

A further ambivalence in current biobank governance lies in the tension between the rhizomic nature of biobank information assemblages and the potential guidance character of 'personalized medicine'. Ideally, one day in the future, the treatment of many diseases will be based on the matching of individualized genomic information and the better understanding of mechanisms of diseases and their treatment. Such treatments might combine elements of health micro-management by states or healthcare maintenance organizations, and aspects of patient choice and self-control. Nevertheless, it is in this future scenario that biobank technologies resemble in many respects also more traditional strategies of biopolitics. While the self-determined individual of modernity remains at the centre of legitimizing biobanks, there can be no doubt that the project of personalized medicine has led to the building of complex infrastructures of body monitoring and surveillance in many countries. While self-guidance by active citizens and patients surely is an option in such still-to-be-realized medical systems, we also can easily imagine more constrained, top-down structured versions of genomics medicine and healthcare dominated by strict regimes of population politics and guided by information flows from biobank projects.

Finally, another ambivalence arises from the potential of biobanks to be fully placed into the service of state power, and the complex liberal-democratic instruments currently used to safeguard the utilization and operation of biobanks. The question needs to be asked whether the 'state of exception' powerfully evoked by Giorgio Agamben in his analysis of sovereignty is not also potentially an integral element of any biobank constellation. There is strong evidence that the complicated systems of anonymization used in all current biobank projects and intended to secure the anonymity of donors seem to have a built-in potential to be put out of order under special circumstances. A prominent recent example is a Swedish biobank for which lawmakers passed a temporary change in the law after the 2004 Tsunami catastrophe giving police the authority to match DNA from bodies in Thailand with blood samples in the biobank, which originally was only intended for medical research. Initially the numbers of the missing and deceased were very imprecise. Nobody knew exactly how many people had died. The only thing certain was that there were many of them. One problem that soon became apparent for the authorities was identifying the deceased. To facilitate identification and thereby assist the close relatives of the deceased, it was decided in parliament to make use of the biobank for this purpose. The PKU biobank was established to detect the metabolic disease Phenylketonuri (PKU), and it contains blood samples of all children born in Sweden after 1 January 1975. The fifth chapter of the Swedish Biobank Law, 'Biobank with specimens from newborn babies', specifically regulates the PKU biobank. The county council of Stockholm is authorized to receive, collect, store, register, and in other ways have at their disposal tissue samples of newborn babies. It states that the parents/guardians must be informed and explicitly grant consent for the submission of a tissue sample from the newborn baby to the PKU register. But on 8 January 2005, in the wake of the Tsunami disaster, the 'Law about Change in the Biobanks in Medical Care Act' (2002: 297) (SFS 2005: 1) was decided by the *Riksdag*. About 240 (out of 349) members of parliament attended. All parties agreed, and the decision was taken without vote. This implied that the biobank could be used for other purposes than it had been intended for, and it also meant that the strict Swedish rules on informed consent were ignored. To this end, a temporary change of the Biobank Law was pushed through at an unusually quick pace. This change met with very little objection, and the debate surrounding it was almost non-existent.[1] In Sweden, we find an interesting combination of evoking a state of exception in the language of assumed informed consent. The state of exception evolved through legal-bioethical procedure. As the Swedish example demonstrates, existing regulatory and ethical safeguards are to a considerable extent provisional and always based on the cooperation of many involved individuals and local contexts.

Conclusions

The great public and scholarly interest in biobanks is strongly connected to the actual and potential value of collections of biological specimens for research, and the possible rise of new forms of medical treatment, such as described in the vision of personalized medicine. But, as this chapter emphasized, biobanks are not only scientific projects and infrastructure developments, they simultaneously constitute a new way of governing life through highly complex social/scientific assemblages consisting of a multiplicity of heterogeneous objects, such as people, signs, chemicals, knowledges, tissues, genes, and institutions that increasingly transcend the boundaries of the nation-state and take on a globalized character. They seem to offer new trajectories of monitoring bodies and are involved in a series of complex transformations of collective identities, healthcare, biopolitics, bio-medical scientific practices, and modes of economic production. Thus, biobanks are new particular strategies through which life is governed.

With biobanks seems to come a new form of the politicization of life, sometimes openly contested, such as in Iceland or Estonia, sometimes 'by stealth', such as in Japan or Sweden, in which the potential 'state of exception' and the potential of the violation of patient and human rights is as much a scenario and topic of discussion as the diligent upholding of principles of bioethics and the new politics of self-guidance in health matters. Thus, contemporary biobank development, one of the major strategies in medical genomics today, emerges as a heterogeneous ensemble that combines 'old' and 'new' modes of biopolitics in a flexible way, adapted to local circumstances and constellations. Pending national elections, tsunamis, the modernization of the healthcare system, economies of hope, or international competition might all be aspects of such contexts that give sense to a new government of life in which the relationship between biobanks, society and the state is at stake, and increasingly become an object of contestation.

Note

1 Research support for the analysis of the changes in the Swedish Biobank Law came from Sarah Kalm.

References

Agamben, G. (1998) *Homo Sacer: Sovereign Power and Bare Life.* Palo Alto, CA: Stanford University Press.

Andrews, L. (2002) 'Genes and patent policy: rethinking intellectual property rights', *Nature Reviews*, October 2002: 804.

Berg, K. (2001) 'DNA sampling and banking in clinical genetics and genetic research', *New Genetics and Society*, 20 (1): 59–68.

Brekke, O. A. and Sirnes, T. (2006) 'Population biobanks: the ethical gravity of informed consent', *BioSocieties*, 1 (4): 385–98.

Brown, N. and Webster, A. (2004) *New Medical Technologies and Society: Reordering Life*. London: Polity Press

Cambon-Thomsen, A. (2004) 'The social and ethical issues of post-genomic human biobanks', *Nature Reviews Genetics*, 5: 866–73.

Crawford, R. (1984) 'A cultural account of "health": control, release, and the social body', in J. McKinlay (ed.), *Issues in the Political Economy of Health Care* (pp. 60–106). New York and London: Tavistock.

Dean, M. (1999) *Governmentality: Power and Rule in Modern Society.* London: Sage.

Deleuze, G. and Guattari, F. (1987) *A Thousand Plateaus.* Minneapolis, MN: University of Minnesota Press.

Foucault, M. (1977) *Discipline and Punish: The Birth of the Prison.* London: Allen Lane.

Foucault, M. (1979) *The History of Sexuality, Vol. 1: An Introduction*. London: Allen Lane.

Haggerty, K. D. and Ericson, R. V. (2000) 'The surveillance assemblage', *British Journal of Sociology*, 51: 605–22.

Jensen, K. and Murray, F. (2005) 'Intellectual property landscape of the human genome', *Science*, 310 (5746): 239–40.

Kay, L.-E. (1993) *The Molecular Vision of Life. Calltech, the Rockefeller Foundation and the Rise of the New Biology*. New York: Oxford University Press.

Knoppers, B. M. and Chadwick, R. (2005) 'Human genetic research: emerging trends in ethics', *Nature Review Genetics*, 6: 75–9.

Marshall, T. H. (1950) *Citizenship and Social Class: And Other Essays*. Cambridge: Cambridge University Press.

Mayrhofer, M. and Prainsack, B. (2007) 'Being a member of the club: the transnational (self-)governance of biobank networks', *International Journal on Risk Assessment and Management* (special issue on 'Inclusive risk governance', edited by S. Funtowicz and M. Craye).

Nelkin, D. and Lindee, M. S. (1995) *The DNA Mystique: The Gene as a Cultural Icon*. New York: W. H. Freeman & Co.

Oxley, H. and MacFarlan, M. (1995) 'Health care reform: controlling spending and increasing efficiency', *OECD Economic Studies*, 24: 7–55.

Patton, P. (1994) 'MetamorphoLogic: bodies and power in *A Thousand Plateaus*', *Journal of the British Society for Phenomenology*, 25: 157–69.

Parthasarathy, S. (2007) *Building Genetic Medicine: Breast Cancer, Technology and the Comparative Politics of Health Care*. Cambridge, MA: MIT Press.

Petryna, A. (2002) *Biological Citizenship: Science and the Politics of Health after Chernobyl*. Princeton, NJ: Princeton University Press.

Plummer, K. (2003) *Intimate Citizenship: Private Decisions and Public Dialogues*. Seattle, WA and London: University of Washington Press.

Rabinow, P. (1999) *French DNA: Trouble in Purgatory*. Chicago, IL: University of Chicago Press.

Rabinow, P. and Dan-Cohen, T. (2005) *A Machine to Make a Future: Biotech Chronicles*. Princeton, NJ: Princeton University Press.

Rose, N. (2001) 'The politics of life itself', *Theory, Culture & Society*, 18: 1–30.

Rose, N. and Novas, C. (2005) 'Biological citizenship', in A. Ong and S. J. Collier (eds), *Global Assemblages. Technology, Politics, and Ethics as Anthropological Problems* (pp. 439–63). London: Blackwell.

Sewell, G. (1998) 'The discipline of teams: the control of team-based industrial work through electronic and peer surveillance', *Administrative Science Quarterly*, 43: 397–428.

Stewart, A. (1999) 'Cost-containment and privatization: an international analysis' in D. Drache and T. Sullivan (eds), *Market Limits in Health Reform. Public Success, Private Failure* (pp. 63–84). London: Routledge.

Waldby, C. (2000). *The Visible Human Project: Informatic Bodies and Posthuman Medicine*. London and New York: Routledge.

Waldby, C. (2002) 'Stem cells, tissue cultures and the production of biovalue', *Health: An Interdisciplinary Journal for the Social Study of Health, Illness and Medicine* 6 (3): 305–23.

Waldby, C. and Mitchell, R. (2005) *Tissue Economies: Blood, Organs and Cell Lines in Late Capitalism*. Durham, NC: Duke University Press.

Part 2

How to build a biobank: comparing different approaches

3 The rise and fall of a biobank

The case of Iceland

Gísli Pálsson

Aristotle partly defined the 'simple' fact of living in opposition to politics. For him, living itself was beyond the activities of the *polis* and, therefore, by definition, outside politics. As Thrift observes, drawing upon Agamben's terminology (1995), one of Foucault's contributions was to shatter this neat distinction, pointing out that, ironically, 'bare life' was increasingly politicized, in fact one of the central concerns of the polis: 'Indeed it is possible to argue that simple natural life is now the most active zone of politics' (Thrift 2004: 147). Biobanks of the kind discussed in this book illuminate the advance and expansion of the realm of biopolitics. While they differ in terms of structure, history, and context, they all draw upon a series of developments in the biopolitical history of states and bureaucracies over the last couple of centuries and their growing concerns with the monitoring and governing of bodies and populations. This chapter outlines the case of the Icelandic Health Sector Database (HSD), exploring some of the critical issues it has raised (see also Pálsson 2007), the main reasons for its termination, and its implications for governance.

In 1998, the Icelandic Parliament ratified a bill on a HSD that would assemble in digital form medical records for the entire Icelandic population. The plan for the database has often been represented as the first population 'biobank' of its kind, a model for others to learn from, to imitate, or to avoid. Since its launching, dozens of somewhat similar plans have been discussed or developed in different parts of the world. While many national, ethnic, and regional biobank projects that have, in one way or another, drawn upon the Icelandic experience are on schedule, surprisingly, perhaps, work on the HSD itself has come to a halt. Given the problems and delays of some of the biobank projects launched at the turn of the century – in particular the Icelandic one – one is entitled to ask: are they simply cases of collapse and the failure of governance?

The terms 'success' and 'failure' – and, indeed, the notions of 'rise and fall' – need some qualification in this context. As Malpas and Wickham point out (1995), positivist social science and twentieth-century Western life more generally tended to concentrate on success and downplay failure, portraying the latter as an aberration or temporary breakdown in the governance of a social system. For them, in contrast, failure is a result of the necessary incompleteness of governance:

> Governance is . . . predicated on the resistance of the object, not only in terms of the being of the object as something recalcitrant, but also in terms of the object as in some sense separate from that which governs. Moreover, since every project is always a part of some more extensive assemblage, so every project is always enmeshed with other projective activities, and there can be no guarantee that such projects, though connected, will even be wholly consistent with one another.
>
> (Malpas and Wickham 1995: 46)

Given such a perspective, a biobank project is necessarily incomplete, a relative failure, embedded in larger projects composed of conflicting as well as complementary components. A 'biobank', then, is not a pre-formed thing readily available for 'experimenting' but an embedded enterprise co-constructed by context. While some narratives suggest the 'shelving' or 'failure' of the Icelandic project was mainly due to legal and ethical developments (Abbot 2004; Winickoff 2006), in reality the reasons were more complex. I argue that there were many reasons for the disintegration of the Icelandic project, financial, managerial, ethical, and political. Above all, the fate of the project underlines problems in governing and the tensions of domestic as well as international bioethics and biopolitics. One of the main reasons for the disintegration of the project, I argue, has to do with what I call the 'biopolitics of the dispossessed', the strategies of resistance developed by segments of the Icelandic medical establishment which suddenly felt deprived of the control and security they had enjoyed in the past, threatened by the biomedical landscape fashioned by the biobank proposal. In the process of adapting to a changing scene the biobank project has been decomposed, redefined, and transformed as part of an emerging biomedical assemblage.

A brief note on the setting is in order. Iceland was probably settled from Scandinavia and the British Isles during the ninth century. Soon after the settlement, a parliament al-thing or *Alþing* was established; thus began the general assembly of Iceland. While the *Alþing* is often seen as one of the earliest forms of democratic national government, this is an exaggeration (Karlsson 2000: 21). The establishment of the *Alþing* signified the beginning of the so-called 'Commonwealth Period' which gave birth to the saga literature, a unique source of information about the stateless social formation in question. Eventually, in 1262, the Commonwealth collapsed due to internal pressure. Iceland remained a colony of Norway and Denmark for centuries, until 1944. Soon after World War II, the economy was modernized and the focus of production shifted from agriculture to fishing. Due to a rapidly expanding fishing fleet, new markets for fish products abroad, and increased fishing effort by foreign fleets on Icelandic grounds, pressure on the main fishing stocks reached critical levels in the 1970s. While fishing no longer has the dominant status it had for years, as the main source of wealth, it has continued to shape political discourse on resource management, governing, and access.

The modern Icelandic state is a parliamentary democracy with a strong welfare system much like those of Scandinavia. Currently, the total population is 300,000. Political culture emphasizes individualism and personal autonomy. At the same time it emphasizes continuity with the past, particularly in terms of language. The Icelandic context, moreover, is characterized by a dynamic entrepreneurial culture. There is a general sense that energetic individuals with creative ideas are able to 'get things done', significantly influencing the turn of events and the shape of economy and society. Entrepreneurial dynamism probably has several roots. For one thing, the small scale of the population provides a sense of relative closeness between the grassroots and the state, commoners and the elite. Also, there is extensive social mobility and cultural homogeneity, in the absence of a traditional elite and rigid class structure. Moreover, since World War II, which brought a new scale of innovation and economic expansion to Iceland, the local population has drawn upon a range of foreign influences, from America, Scandinavia, and the rest of Europe. This mixing of different traditions in many fields, including education, research, banking, technological developments, and industries, has created a fertile soil for entrepreneurial culture.

The health sector database

The plan for the construction of a Health Sector Database on the Icelandic population was initiated by deCODE genetics, a company established in 1996 by two physicians, the Icelander Kári Stefánsson (a Professor at Harvard University) and his collaborator Jeff Gulcher. The plan was to use the relatively homogeneous Icelandic genome and the wealth of local historical records for the purpose of advancing biomedical research on the genetics of common diseases.[1] Such an initiative has to be partly seen within the context of international developments at the turn of the century when genomics and associated technologies and bioindustries were being developed at an exponential rate, partly as a result of spectacular advances in the understanding of life and partly as a result of promises of a new gold rush. The Human Genome Project had already been launched, signalling scientific breakthroughs and a new scale of genomic enterprises. The time seemed to be ripe for large-scale enterprises bringing the dynamics of the market into population genetics, in some kind of partnership with national and international governing bodies. Not only were genomic enterprises shaped by the international scientific and economic context, also the rules of the game were increasingly being framed within the global governance and politics of bioethics. As it unfolded and eventually disintegrated, the HSD project was both informed by the global context and instrumental in its development. The local and the global mutually constituted each other.

Immediately after the presentation of the first bill for the Health Sector Database to the Icelandic Parliament, there was a strong public reaction both domestically and internationally. After nine months of national debate, in December 1998 the Icelandic Alþingi passed a bill authorizing the construction of the database (Icelandic Parliament 1998, Act no. 139). Given the pioneering nature of the enterprise, a legal and ethical model or framework for such a comprehensive, national project was nonexistent. The license to construct the biobank was open to competition; the licensee would finance it, and the end result would belong to the National Health Service, with the licensee retaining privileged rights to commercialize it for twelve years. deCODE genetics was granted the license to construct the database. The text for the license specified that the company should pay the Icelandic state a certain fee for the assembly and use of the records of the medical service. deCODE genetics was requested to cover the cost of the agreement on the database, its construction, and marketing. In addition, it had to forfeit 70 million Icelandic kronur annually (about 1 million $US) for the license that would be used for furthering medical research and development. Furthermore, the Icelandic state would receive 6 per cent of the annual profit that deCODE genetics would make from using the database.

The public debate in Iceland frequently referred simply to 'the database issue' (*gagnagrunnsmálið*), subsuming medical records (the HSD), genetic information, and genealogies as well as their combination (see Figure 3.1). International discussion also emphasized the combination of these three datasets and their collective commodification. This is not surprising as original plans suggested combining medical records and the other two databases for scientific and commercial purposes, thereby developing a 'genetic database'. The legal framework, however, of the Health Sector Database is exclusively focused on the assembly of medical records. Genetic samples would be collected for specific research purposes and only combined with medical records on the condition of scientific and ethical screening and approval.

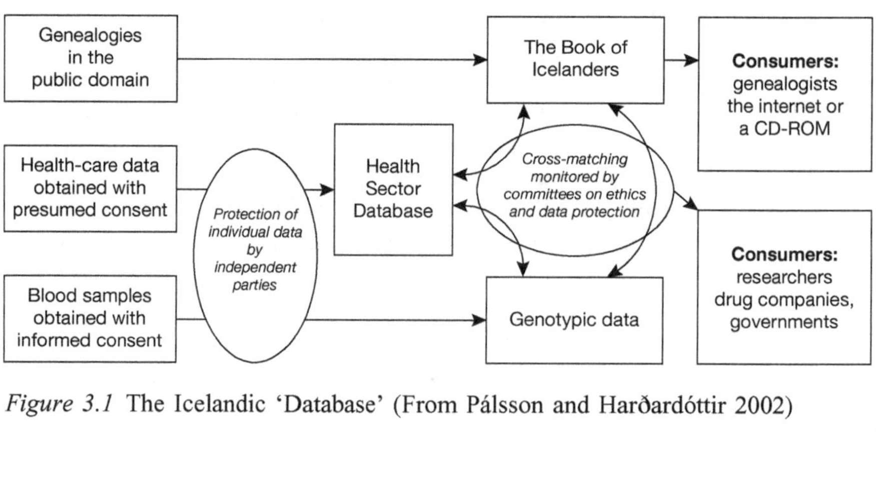

Figure 3.1 The Icelandic 'Database' (From Pálsson and Harðardóttir 2002)

For the spokespersons of the HSD, the power of genetic and epidemiological analyses would be greatly enhanced by the project. The medical records available since 1915, it was argued, would allow for the exploration of a set of new questions on the interaction among a number of variables apart from genetic makeup and genealogical connections, including variables pertaining to lifestyle, physical and social environments, the use of particular medicine, and degree and kind of hospitalization. A relatively homogenous population with good genealogical as well as medical records would be the ideal experimental site for biomedical analysis. According to deCODE genetics and their investors, the Icelandic context met such conditions better than most other contexts. No doubt, deCODE's early statements about the value of biomedical research in the Icelandic context overemphasized the genetic homogeneity of Icelanders and their isolation through much of their history, partly to appeal to research funds, policy makers, and investors. Genetic analysis, however, shows that the rhetoric has more than a kernel of truth.[2]

There were several other kinds of motives than scientific advantage for the establishment of the HSD. While Icelandic healthcare is highly efficient, and Iceland ranks at the top internationally in terms of several measures, including longevity, it is quite expensive. From the point of view of members of Parliament, ministers and representatives of the state, the HSD would render the healthcare system more effective. Also, the database would provide a range of new jobs and firmly place Iceland in the growing new economy, an important asset in an economy heavily dependent on one kind of overexploited resource, the fisheries. From the point of view of the Icelandic medical authorities and spokespersons for deCODE genetics, the HSD, then, had clear scientific and managerial motives.

Almost from the beginning, the public supported deCODE genetics and the database project. In April 2000 a Gallup survey concluded that no less than 81 per cent of Icelanders supported the database, while 9 per cent were opposed and 10 per cent were neither for nor against. One further sign of public support for the HSD was the strong positive initial response of the Icelandic stock market to deCODE genetics, although the reaction of the market later on has been mixed and shifting. deCODE genetics and the database project, supporters of the project argued, would advance biomedical research in Iceland, creating many new positions for scientists and laboratory assistants both within the company and at the University of Iceland. deCODE genetics, it was often pointed out, had attracted much investment from abroad and created numerous jobs for Icelandic scientists, many of whom in the absence of deCODE genetics would have been forced to seek employment abroad. The arguments of the proponents of the project also tended to emphasize the opportunities it provided in terms of private initiative. Furthermore, local interest in the database has to be partly seen within the context of 'Norse' history. The nationalist discourse of Icelanders – with its emphasis on the sagas, the glories of the past, and the 'uniqueness' of Icelandic heritage – partly explains the fascination of modern Icelanders with genetic databases and family histories.

A strong ethical and political body – *Mannvernd*, the Association of Icelanders for Ethics in Science and Medicine – was formed in direct response to the database project. Its main spokespersons were physicians, biologists, geneticists, and philosophers. Analysis of newspaper articles mentioning the database project published in the heat of the debate shows that physicians wrote no fewer than 28 per cent of the articles – understandably, perhaps, since they had compiled many of the medical records that are the subject of debate (Pálsson and Harðardóttir 2002). Their contributions turned out to be overwhelmingly against the project. The opposition to the project emphasized ethical concerns, particularly that of privacy, the protection of personal medical information. Another important concern for much of the opposition was that of informed consent. The HSD was to operate on the basis of the principle of *presumed* rather than informed consent; people could refuse to be included in the collective medical records, but if they did not, information on them would be automatically entered.

Much of the discussion of the database project focused on issues of property and control. Thus, a fundamental debate took place concerning the ownership of and access to genetic information and medical records. Perhaps the dominant focus in the debate was the fact that a private multi-national company proposed to explore the genetic bases of common diseases in the entire Icelandic population and to commercialize its results. Some of the critical newspaper comments in this genre echoed claims about 'biopiracy' popular in debates on genetic research on indigenous groups and the Human Genome Diversity Project. The property issue has often been discussed with reference or in comparison to ongoing debates about another thorny common-property issue, namely, the allocation of individual transferable quotas to rights in fish. Medical (and possibly, genetic) information, it was argued, are common-pool resources with some of the characteristics of the fishing stocks in Icelandic waters. Privileged access, permanent or temporary, should therefore be granted only in return for a fee to ensure equity and fairness. For many of the critics, however, the commodification of biomedical data was problematic.

The heat of the opposition to the project was driven by an apparent sense among physicians of rather suddenly lacking authority in the biomedical domain, of losing dominion over information largely constructed and controlled by them in the past. Some academics alleged that the restrictions of access to information and resources implied in the privileged contract of deCODE genetics with the Icelandic state would inevitably result in the stagnation of bioscience. Thus, the sub-text of some of the debates centered on where cutting-edge research on the Icelandic human genome occurred and where it should be located in the future. With the construction of the database, physicians and academe were being removed from the discursive center of local biomedicine, making way for state officials, database staff, and spokespersons for deCODE genetics.

The biopolitics of the dispossessed

When it was launched, the HSD was presented as a fundamentally new avenue to biomedicine, drawing upon the successes of the new genetics and the dynamics of the market. About twenty-five people worked full-time on the project for over two years doing preparatory work. Precursory computer programs were developed and tested. A prototype was prepared, a 'container' into which the data would be fed once everything was clear. In addition, a pilot project involving the healthcare institute of the town of Húsavík in northern Iceland was set up to explore some of the problems involved. Moreover, ethical and legal frameworks were discussed and revised. In the process of preparing all of this, deCODE staff would tour the country, visit health clinics, and introduce their project to local leaders. More than fifty institutions throughout the country had to be consulted. Despite all of this diplomatic effort, the project came to a halt. Most of the key people associated with it seem to assume that it is no longer on the agenda although the death certificate has so far not been issued in public. Given the buzz surrounding the plan for the biobank, the first of its kind, one is entitled to ask why it has not materialized.

Interviews with some of the people working on the construction of the database revealed important reasons and concerns. Several kinds of problems emerged and caused delay. Some of the difficulties encountered by the deCODE team were purely technical ones. Digitalizing and assembling massive amounts of data from different periods and different healthcare institutes demanded extensive work on software development and the standardizing of criteria to be applied in the entry of data. This probably proved more complex and time consuming than anticipated. Relations with the local communities involved were also problematic at times. In many cases, there were demands for some kind of rewards, in particular new jobs, involving local people in the digitalization of medical records, a massive corpus of information that had accumulated in paper form for decades.

A further reason for slow-down relates to what might be identified as the biopolitics of the dispossessed. The HSD would only become useful if records from local clinics were passed on to the licensee for assembly. The Icelandic Medical Association, however, claimed from the beginning that the database would violate the relation of trust between physician and patient, since it would operate on the basis of the principle of presumed consent, thereby violating the standard practice of informed consent. As the database was under construction, some directors of local clinics refused to hand over 'their' records, emphasizing their responsibility to their patients and pointing out that the latter had not consented to any such transfer of information about them. The main organized opposition to the project, the Association of Icelanders for Ethics in Science and Medicine, supported their refusal to hand over local records. Some patient groups complained that the doctors were not consulting patients and that they had no right to make decisions about medical records on their behalf. In effect, they argued, the doctors were claiming ownership and control of information that properly belonged to both the patients themselves and the community – the Icelandic state that had financed their recording and assembly.

Inevitably, the opposition of the Medical Association slowed things down. While many of the necessary preparations for the HSD had been made, contracts with local clinics were not signed because of professional resistance. Sometimes the only local physician was passionately opposed to the project and local people felt obliged to proceed carefully and to avoid confronting them, commenting 'they didn't want to lose their doctor'. In some cases, medical data were not passed on despite the fact that clinics had signed the relevant agreement. Sometimes, the records for an entire town had to be 'abandoned' due to physicians' resistance; one key physician in a fishing community insisted that 'his' data would not be transferred: 'over my dead body!' In another case, every time the deCODE staff would arrive for discussions the local physician would disappear from the scene with some kind of excuse. While the preparation and handing over of local health records would present some burden to local clinics in terms of time and resources, deCODE staff were taken by surprise by the resistance. Why would local data not be almost automatically passed on to the makers of the database, given that the authorities in question, Parliament and the Ministry for Health, had decide to get this done and to offer the licence and the responsibility to deCODE?

One stumbling block in the making of the database project was the fact that a growing number of people opted out of it, refusing to pass on their personal information. By June 2003, roughly 20,000 people had opted out, a significant figure given the size of the population. A further setback was a decision by the Supreme Court in November 2003. The case, *Ms. Ragnhildur Guðmundsdóttir vs. the Icelandic State*, centered on the legality of presumed consent with respect to medical information regarding children, incompetent adults, and the deceased (Supreme Court of Iceland 2003, no. 151). Ms. Guðmundsdóttir protested against the transfer of data pertaining to her deceased father to the database. Would it be meaningful to apply the principle of presumed consent to people who were not in a position to opt out of the database? The Court acknowledged the rights of relatives of deceased persons to make decisions about the data involved, thereby adding one more complication to the database project.

An important problem related to the security targets set by the Office for Personal Data Protection. deCODE staff suggest the targets set for protecting the anonymity of samples and data were both too high and too cumbersome to work with. A tight 'wall' or 'curtain' was established between the researcher and the data for protection against potential 'malicious users', making meaningful work on the data nearly impossible. This was partly the result of miscommunication between two groups with rather different training and perspectives: deCODE scientists and state lawyers. The Office for Personal

Data Protection suggested a 'veil' or 'curtain' between the researchers and the raw medical data. This put heavy constraints on the project. As one of the deCODE staff commented:

> Usually when I work with health records . . . I operate with some kind of matrix with a host of variables such as age, the use of medicine, blood pressure etc. I can scan all the lines and quickly spot some of the obvious errors there may be. Now I couldn't do this. The staff of the Office for Personal Data Protection were so scared of the possibility that we might recognize the individuals in question. They felt they had to change the averages a bit by means of some kind of algorithm; otherwise, they reasoned, everything might be traceable. As a result, we had to fumble about in the dark.

Negotiations about security targets got increasingly strained. The running of the HSD could only begin once the Office would be satisfied with the security measures employed. deCODE complained with the analogy of building a hydroelectric power plant, a pertinent analogy in the Icelandic context. Would it make sense to invest in a massive project like that under the uncertainty that for some reasons channelling water to the turbines might not be permitted once the plant had been completed? No one, it was reasoned, would agree to take such risks. In effect, the Office for Personal Data Protection functioned as an ethical supreme court, with a final word in the screening process and no possibility of appeal.

The Icelandic debate was not a closed national case. The international press, the transnational scientific community, and the emerging informal, international network of bioethicists were heavily focused on the HSD. Part of the reason why the Icelandic case was frequently reported had something to do with the somewhat risky but skilful handling by deCODE genetics of public relations, including its frequent reference to genetic 'roots' and the 'Viking' past. Icelanders were pictured as the guinea pigs of the new genetics, equivalent to model organisms such as the fruit fly. 'Expert' witnesses from the transnational world of science and bioethics often addressed the key issues internationally at conferences, in the media, and on the web (see, for instance, Rose 2003). The Icelandic plans were fiercely opposed, usually from a bioethical vantage point emphasizing patients' rights, informed consent, and the protection of privacy. The response of the international press and academe partly reflected a growing competition between similar projects on the global scene. Recently, Francis Collins, the director of the US National Human Genome Research Institute, suggested, drawing attention to biobank projects proposed or underway in the United Kingdom, Iceland, Estonia, and Japan, that the United States could ill afford not to invest in its own population-based cohort study.

An economic backlash experienced by deCODE at a critical moment had important implications for the database project. The company was forced to make about one-third of its personnel redundant. In an attempt to make ends meet, it streamlined its research on patient groups and increasingly moved into drug discovery and development. At the same time, work on the HSD project was slowed down. The financial backlash at deCODE was partly the result of an international backlash in bioindustries which, in turn, was partly a consequence of debates about the human genome projects. For some time there was an intense tug-of-war between the communitarian and private perspectives represented by the public genome project and the company Celera. Did the human genome belong to humanity or was it up for enclosure by private interests and industry much like the high seas many years ago?[3] Billions of dollars were drained from the biotech sector, more than $2 billion from Celera alone (Shreeve 2004: 324). Stock prizes on the Nasdaq exchange collapsed, with serious consequences for many biotech companies throughout the globe. Some genetic database projects were slowed down or scrapped, testifying to the global nature and implications of developments in biotechnology and biomedicine.

Some of those interviewed experienced a growing disillusion inside deCODE genetics with the database project. One scientist, a former employee of the company commented: 'I was hired when the project began. Then, in 2002 I sensed that something was brewing. I lost faith in what I was doing, thinking "this will never be realized". I was moved to another task, to a patient-group project. This was what the company got payments for. Soon after, I applied for a job elsewhere.' Faced with increasingly winding and time-consuming negotiations and legal battles, rising costs, and growing economic difficulties in the biotech industry, deCODE genetics may have decided to back off and resort to an alternative strategy, expanding its own internal patient-group database and escalating its work on drug development. Currently, the company has acquired genetic samples from about half of the adult Icelandic population through its work on specific common diseases. Despite the shelving of the database project, it has made important advances in genetic research through conventional avenues more or less independent of the biobank enterprise. Perhaps the most impressive achievement of deCODE is a recent clinical trial involving the risk of myocardial infarction. An editorial in the *Journal of the American Medical Association* suggested that while the trial, 'one of the first human trials to attempt to translate genomic findings into clinical practice for cardiovascular disease', should be viewed as preliminary it 'provides an exciting attempt to translate genetic findings to clinical applications' (O'Donnell 2005: 2278). Results like these may have important practical implications, paving the way to diagnostic tests to identify people carrying the variant genes, preventive measures in terms of lifestyle, gene therapy (a particularly controversial procedure), and, possibly, the development of drugs that affect expressions of the proteins involved.

The health authorities and the state may also have decided to define their own alternatives to the HSD although officially they may still be committed to it, developing scattered databases and collections which might later on be somehow combined. From now on, medical records are stored in digital form in the clinics of the National Health Service where they materialize, forming a potential future database that might eventually be combined with the records of the past. Also, a centralized database on medical prescriptions and drug use is already in existence. Finally, one of the spin-offs of the database project is the genealogical Book of Icelanders, which has proved quite useful in deCODE's research on patient groups.

Discussion

While biobanks differ in several respects, they all represent the extension of the biomedical gaze. As Lyotard points out, data banks are 'the Encyclopedia of tomorrow. They transcend the capacity of each of their users. They are "nature" for postmodern man' (1984: 51). Some of the important early biopolitical developments that pre-date biobanks relate to practices of writing, tabulating, and computing. In Medieval Europe, bioinformatics focused on documenting births, deaths, and marriages. During the nineteenth century, statistics and probabilistic methods were developed to describe the health risks and tendencies of national populations and their subdivisions. Adopted in one nation state after another, they represented concerns with keeping track of the health of the general public and governing the national body. Hacking suggests, however, that the 'avalanche in numbers' that was generated in the process was rarely efficient in managing the population of study: 'The fetishistic collection of overt statistical data about populations has as its motto "information and control," but it would more truly be "disinformation and mismanagement"' (1982: 280). In his view, on the other hand, the new 'statistics of sickness' had subversive effects, namely 'to create new categories into which people had to fall, and so to create and to render rigid new conceptualizations of the human being' (1982: 281). While the eighteenth and nineteenth centuries gave birth to biopolitics, the politicization of bare life was taken to its extremes during the twentieth century.[4] The dark shadow of concentration camps and the eugenics that often comes with them continues to inform, and to misinform, discussions of biopolitics, including biobanks.

Hacking's argument (1982) about the fetishism of numbers and disinformation in the collection of statistical data about populations may not be entirely valid for the kind of population biobanks discussed here. After all, they are likely to be efficient biomedical tools, speeding up analyses of the distribution and causes of common diseases. Nevertheless, Hacking's point draws attention to the different and somewhat contradictory agendas of biopolitics (of governments, politicians, companies, etc.), unexpected developments, redefined agendas, indirect spin-offs, and the potential clash between promises and results characteristic for the genome era (see, for instance, Taussig 2005, Lock 2005). In a sense, the Health Sector Database has been decomposed, much like a fragmented human body, recombining with other projects serving different times and agendas. Quite possibly, in the future large-scale genetic databases will be operative at numerous sites, either permanently or from time to time. For gaining larger numbers and samples, which may be important for analyzing many common diseases, different database projects may find it feasible to temporarily share data under some kind of permanent institutional umbrella.

Database projects are not just innovative technical, scientific, and economic enterprises; they are radical experiments in biopolitics with potentially diverse social implications for individuals, families, communities, and the public at large. The broad responses by the international community to such experiments, however, vary from one case to another (Pálsson and Rabinow 2005). No doubt, there are several reasons why the Icelandic case was particularly contested. One important reason probably relates to the fact that the population of an entire nation was involved, not a sample. Many critics found such an inclusive approach shocking, evoking an image of a totalitarian, if not eugenic, regime. I have argued that the main reason for the disintegration of the HSD project has to do with the resistance of local physicians, the biopolitics of the dispossessed. This, in turn, was heavily informed by bioethics on both the local and the global scene, as it grappled to deal with the developments of the genome era partly through its discourse on informed consent.

The rapid expansion of bioethics during the second half of the last century need not be surprising. By definition, the emergence of the new genetics invited concerns with both individual liberties and the collective rights and responsibilities of nations, populations, and ethnic groups. A growing governmentality literature explores how certain phenomena become formulated as problems, 'investigating the sites where these problems are given form and the various authorities accountable for vocalizing them' (Xavier Inda 2005: 8).[5] Interestingly, in the past decade or so, bioethics seems to have taken a communitarian turn, shifting from the perspective of the autonomous individual and the heavy reliance on informed consent which it has advocated in the recent past to a more holistic perspective, emphasizing the principles of reciprocity, mutuality, solidarity, citizenry, and universality. Knoppers and Chadwick suggest that a new 'participatory approach' has emerged as a result of the growing influence of social science on ethics and the reinterpretation of the concept of 'expertise' in genetic ethics. 'There might not', they conclude, 'and cannot, be universal norms in bioethics, as emerging ethical norms are as "epigenetic" as the science they circumscribe' (2005: 78). Thus the principle of informed consent that was at the centre of the storm around genetic databases early on has been losing ground. Some ethnographic evidence indicates that bioethical practice may not have the same meaning to bioethicists and the 'lay' people they claim to speak for. Thus, Almarsdóttir *et al.* (2004) suggest, on the basis of focus group discussions in the Icelandic context, that the issues of confidentiality, privacy, and data protection tend to be framed in a fashion that is rather different from the perspective of personal autonomy prescribed by orthodox bioethics.

It is notoriously difficult to draw lessons from a single case. While Iceland is a welfare state that shares many of the characteristics of other contexts, including Scandinavia, the Icelandic context is somewhat unique, with its nationalistic rhetoric and its emphasis on ancient roots and entrepreneurial dynamism. Each biobank enterprise has a path-dependency of its own, defined by the initial conditions of its making and the evolving context of which it is a part. Had the Icelandic project been launched from the beginning as a truly collaborative venture between academe, government and industry, its life course might have been entirely different – and the strategies of resistance might not have emerged or taken a radically different shape. Had the original plan underlined informed rather than presumed consent, the concern with security targets would perhaps have been less overwhelming and the slow-down, perhaps, would have been less effective. Had the 'communitarian turn' in bioethics taken place prior to the launching of the HSD, the Icelandic controversy might never have taken off. Had the economic backlash experienced by the biotech industry occurred a few years later, deCODE might not have shifted its course. And so on. Historical speculations along these lines, with unreal past conditional clauses, are not particularly fruitful. In the unfolding of the Icelandic case, the politics of biobanking and the HSD were co-constructed, mutually informing each other and similar biobank enterprises later developed in other contexts. As Winickoff observes, the Icelandic Health Sector Database became an experimental site for genomics and genomic governance in the sense that it 'helped produce the technological, political, and normative terrain of all large-scale genomics initiatives today, not just Iceland's' (2006: 97).

Acknowledgements

The research on which this article is based was generously funded by the Nordic Committee for Social Science Research (NOS-S), the Icelandic Center for Research (Rannís), and the Research Fund of the University of Iceland. I thank Dr Anna Birna Almarsdóttir, Valgerður Gunnarsdóttir, and Kristín Erla Harðardóttir for their suggestions and remarks regarding some of the issues discussed here. Also, I appreciate comments on an early draft by other contributors to this book, in particular the editors.

Notes

1 The idea of constructing a comprehensive population database on Icelanders was not a novel one. In an essay written in 1943, the novelist Halldór Laxness suggested that an 'anthropological' office or institute (*mannfræðistofnun*) be established in order to document, for the purpose of marketing and research, information on every Icelander ever recorded: 'Information on every family would be organized (*kerfaðar*) so that the employees of the institute would be able to assemble, at short notice, the family history of any Icelander . . .' (Laxness 1962: 155–6). A precursor to the Health Sector Database was a database made from the early

1960s onward under the umbrella of the University of Iceland and largely funded by the Atomic Energy Commission of the United States. In 1966, a Genetic Committee was established at the University of Iceland, the role of which was to encourage and organize genetic research at the University and, more generally, in Iceland (see Pálsson 2007).
2. A recent study, by deCODE researchers, suggests the 'overwhelming conclusion . . . is that the Icelandic gene pool is less heterogeneous than that of most other European populations' (Helgason *et al.* 2003: 283). A review of the literature of human genetic diversity in Europe supports such a conclusion, arguing that 'Icelanders do show evidence of greater drift effects than most other European populations' (Barbujani and Goldstein 2004: 138).
3. Attempts were being made to reconcile the plans and interests of the two projects, but negotiations repeatedly collapsed. In March 2000, US President Bill Clinton and Prime Minister of Great Britain Tony Blair issued a joint statement in an attempt to bring things under control, urging nations, scientists, and corporations to freely share their information. The effect of this statement was immense.
4. The extreme case is represented by the formation of concentration camps by the German Social-Democratic government in the 1920s and later on during the Nazi regime. For Agamben, the camp is 'the hidden matrix and *nomos* of the political space in which we are still living' (Agamben 1995: 166), a space characterized by the politicization of bare life, the new biopolitical body of humanity.
5. Maskell and Pelts criticize the conception of ethics that 'turns the ethical code into a kind of "constitution" . . . and the professional into an adjudicator who, on the basis of this ethical constitution and his mastery of expert information, assumes a position of unquestioned (and often implicit) superiority' (2005: 3). For them, it is essential to 'embed' ethics, to locate ethics not in a law-like universal or a ready-made scheme but in practices of interaction.

References

Abbot, A. (2004) 'Icelandic database shelved as court judges in peril', *Nature*, 429 (13 May): 118.

Agamben, G. (1995) *Homo Sacer: Sovereign Power and Bare Life*. Trans. D. Heller-Roazen. Stanford, CA: Stanford University Press.

Almarsdóttir, A. B., Traulsen, J. M. and Björnsdóttir, I. (2004) '"We don't have that many secrets": the lay perspective on privacy and genetic data', in G. Árnason, S. Nordal and V. Árnason (eds), *Blood and Data: Ethical, Legal and Social Aspects of Human Genetic Databases*. Reykjavík: University of Iceland Press & Centre for Ethics.

Barbujani, G. and Goldstein, D. B. (2004) 'Africans and Asians abroad: genetic diversity in Europe', *Annual Review of Genomics and Human Genetics*, 5: 119–50.

Hacking, I. (1982) 'Biopower and the avalanche of printed numbers', *Humanities in Society*, 5: 279–95.

Helgason, A., Nicholson, G., Stefánsson, K. and Donnelly, P. (2003) 'A reassessment of genetic diversity in Icelanders: strong evidence from multiple loci relative homogeneity caused by genetic drift', *Annals of Human Genetics*, 67: 281–97.

Icelandic Parliament (1998) Act on a Health Sector Database no. 139. http://eng.heilbrigdisraduneyti.is/laws-and-regulations/nr/659#allt.

Karlsson, G. (2000) *Iceland's 1100 Years: History of a Marginal Society*. London: C. Hurst & Co.

Knoppers, B. M. and Chadwick, R. (2005) 'Human genetic research: emerging trends in ethics', *Nature Review Genetics*, 6 (January): 75–9.

Laxness, H. K. (1962 [1943]) 'Mannlíf á spjaldskrá' ('Indexing human life'), in H. K. Laxness, *Sjálfsagðir hlutir: Ritgerðir*. Reykjavík: Helgafell.

Lock, M. (2005) 'Eclypse of the gene and the return of divination', *Current Anthropology*, 46 (supplement): 47–70.

Lyotard, J.-F. (1984) *The Postmodern Condition: A Report on Knowledge*. Minneapolis, MN: University of Minnesota Press. Translated from the French by G. Bennington and B. Massumi.

Malpas, J. and Wickham, G. (1995) 'Governance and failure: on the limits of sociology', *Australian and New Zealand Journal of Sociology*, 31 (3): 37–50.

Maskell, L. and Pelts, P. (eds) (2005) *Embedding Ethics*. Oxford: Berg.

Nationaler Ethikrat (2004) *Biobanken für die Forschung Stellungnahme*. Berlin: Nationaler Ethikrat.

O'Donnell, C. J. (2005) 'Translating the Human Genome Project into prevention of myocardial infarction and stroke – getting close?', *Journal of the American Medical Association*, 293: 2277–9.

Pálsson, G. (2007) *Anthropology and the New Genetics*. Cambridge: Cambridge University Press.

Pálsson, G. and Harðardóttir, K. E. (2002) 'For whom the cell tolls: debates about biomedicine', *Current Anthropology*, 43 (2): 271–301.

Pálsson, G. and Rabinow, P. (2005) 'The Iceland controversy: reflections on the trans-national market of civic virtue', in A. Ong and S. J. Collier (eds), *Global Assemblages: Technology, Politics, and Ethics as Anthropological Problems*. Malden, MA: Blackwell Publishing.

Rose, H. (2003) 'The commodification of virtual reality: the Icelandic health sector database', in A. Goodman, D. Heath and S. Lindee (eds), *Genetic Nature/Culture*. Los Angeles, CA: University of California Press.

Shreeve, J. (2004) *The Genome War: How Craig Venter Tried to Capture the Code of Life*. New York: Alfred A. Knopf.

Supreme Court of Iceland (2003) Ragnhildur Guðmundsdóttir v. the Icelandic State, no. 151.

Taussig, K.-S. (2005) 'The molecular revolution in medicine: promise, reality, and social organization', in S. McKinnon and S. Silverman (eds), *Complexities: Beyond Nature and Nurture*. Chicago, IL: University of Chicago Press.

Thrift, N. (2004) 'Bare life', in H. Thomas and J. Ahmed (eds), *Cultural Bodies: Ethnography and Theory*. Oxford: Blackwell Publishing.

Winickoff, D. E. (2006) 'Genome and nation: Iceland's health sector database and its legacy', *Innovations: Technology, Governance, Globalization*, 1 (2): 80–105.

Xavier Inda, J. (2005) 'Analysis of the modern: an introduction', in J. Xavier Inda (ed.), *Anthropologies of Modernity: Foucault, Governmentality, and Life Politics*. Oxford: Blackwell.

4 Estonia

Ups and downs of a biobank project

Rain Eensaar

Introduction

Unlike the Icelandic biobank project, which gained international attention since it was first presented to the world, the Estonian Genome Project (EGP) has never achieved similar interest. But despite its notoriety, the Icelandic project failed, while the Estonian project is still alive and well after a difficult time in early 2000, and it is again gaining momentum. The Estonian biobank project provides a fascinating example of the difficult process of building and governing a biobank. In this chapter we will trace the challenging path of the EGP, and how it was woven and negotiated into the complex Estonian political, economic, social, and cultural fabric.

On the most general level, the EGP aims to establish a database that compiles phenotype and genotype data of a large portion of the Estonian population. The project has experienced highs and lows during its short history. Although the scientific organization and the concept of the project were working well, the initially promising public–private partnership failed after three years of venture capital financing over 2001–3. When at the end of 2003 private investors started to postpone payments and discuss changing objectives, the project almost collapsed because of a lack of available funding. At the end of 2005 government ministries finally decided to provide public funding for the necessary continuation of the project. Today the path is cleared for the EGP, and it is planned that large-scale data collection, which started in 2007, will increase ten-fold from its current 13,500 samples to 100,000 samples by 2010.

From a governance perspective, four dimensions are of special interest about the EGP: first, the Estonian biobank project was presented to the public as a 'bridge to Europe', an opportunity to demonstrate the quality of Estonian science in the European context, and the role of this framing in the legitimization of the biobank project is noteworthy. As we shall see, much effort was spent contextualizing the Estonian biobank project within policy narratives dominant in Estonia, thus linking the biobank to Estonian political identity. This explicit linking of the Estonian Genome project to national identity is not unique in comparison, as for example the case of the Western Australian Genome project shows (see McNamara and Petersen, Chapter 12), but it nevertheless is an interesting feature in making Estonian genome research governable. Second, the Estonian biobank project also initiated a dialogue with 'the public' in an attempt to communicate its goals and ambitions to larger audiences. The decision to create special legislation and public education programmes can be seen as deliberate efforts to mediate between science and society, and to create legitimacy for the project. Third, less articulated, but clearly present, was the specific health governance dimension of the project, and at least an implicit ambition to contribute to a transformation of healthcare in Estonia through actively reshaping the interaction between patients and the health system by trying to build a new system of personalized medicine. Whereas the focus of many biobank projects is mainly on research, the Estonian project, at least for the time being, also has a strong applied dimension and the ambition to transform the national healthcare system, which is shared with the UK Biobank project (see Corrigan and Petersen, Chapter 9) and Biobank Japan (see Triendl and Gottweis, Chapter 8). Finally, the Estonian biobank project highlights the essential financing component of biobanking: the near financial collapse of the project is an episode that demonstrated that a viable business plan and sustainable financing are necessary key elements in biobank governance, and that financing through commercial sources can quickly derail a particular biobank project, a point that the Icelandic case seems to highlight (see Pálsson, Chapter 3).

EGP: the story unfolds

As mentioned, initially the Estonian Genome Project was presented as a project with a very special national dimension. In this context, it must be remembered that the political landscape and civic culture of present-day Estonia have been influenced both by the short duration of democratic culture in 1918–39 during the first independence period, and by political corporatism since 1991, a society inherited from late totalitarianism and recent rapid socioeconomic changes that damaged all sectors and layers of society. With a population of about 1.3 million citizens and an area of 45,227 square kilometers, it is remarkable that Estonia had survived centuries of foreign rule to emerge as an independent state for the second time in 1991 (Unwin 1999: 165–6). The restoration of Estonian political identity unfolded in different fields, of which the most important were the 'Estonisation of Estonia', the demarcation of the Estonian state, and the 'return to Europe'. Whereas 'Estonisation' and the demarcation of the Estonian state served primarily to separate Estonia from Russian spheres of influence and also dealt with the Estonian Russian population, the 'return to Europe' discourse outlined a broad scenario for Estonia's path into the future. Science, technology, and innovation were instrumental in this context (Aalto 2001: 110).

When the possibility of an Estonian biobank was first considered, the project quickly came to be associated with 'Estonia's return to Europe'. As one of our interview partners, the leader of the Estonian biobank project expressed it, the project was seen as a 'way to Europe' (Interview Metspalu, 5 April 2005). The EGP was also framed as becoming potentially the 'Estonian Nokia'. In a May 1999 speech delivered to a business audience, former Estonian president Lennart Meri argued that Estonia should take its cue from Nokia, Finland's powerful telecommunications firm, an idea that was highly popular in the local innovation discourse. In Meri's words:

> With great persistence, I have kept asking what is the Estonian Nokia. Finally, after many months, this question has also reached the pages of our popular media. It has become part of the Estonian folklore. As a writer, I am of course overjoyed to see that my work is alive and creating the silverwhite frame of mind. The conclusion of many Estonians – including several prominent figures – that the President is looking for a single product just as Pippi Longstocking once looked for Spunk – is still somewhat unexpected. The Estonian President is not supposed to seek the Estonian Nokia. I am seeking it for you. For your lazy minds. Every Estonian proprietor must seek it for himself, must seek at least six Nokias every year.
>
> (Meri 1999)

The idea to move with a comparatively large technology project boldly into a new field also resonated with the popular policy idea of the 'Tiger Leap'. For example, Estonia had played a pioneering role internationally in e-government and e-elections. The goal of the Tiger Leap programme was to foster the innovative spirit in Estonia with a national programme whose overall objective was to promote the Estonian educational system by introducing modern information and communication technology (Tiger Leap Foundation). Taking advantage of the biotech revolution was regarded as another window of opportunity in addition to the already favourable image of an efficient IT-developed country with highly respected economic freedom and with the best per capita foreign investment in the region.

With these rhetorical moves representing the Estonian biobank project, the idea of the biobank had been positioned as a policy agenda that could play a key role in considerably strengthening the Estonian economy, creating an Estonian biotechnology industry, and helping Estonia's 'return back' to Europe. Thus, the Estonian Genome project was not only framed as a key project for the general economic position of Estonia, but as an active contribution in the reshaping of the new, democratic, post-Soviet Estonia.

Planning the project

Like many important biobank projects, the Estonian project started as an initiative of a maverick scientist, Andres Metspalu, professor of biotechnology at the University of Tartu. Under his leadership, a group of experts presented the project in 1999 as an unprecedented scientific effort that could help put Estonia back on the map of international science and possibly create worldwide interest. Besides its scientific goals, the project was framed as a chance to increase cooperation with the international scientific community, to boost Estonian biotechnology, and have direct impact on healthcare (Interview Metspalu, 5 April 2005). During this time the government, headed by Prime Minister Mart Laar, had been actively looking for innovative ideas and was therefore eager to support the project.

A central argument supporting the project was that it would create a valuable resource for making world-class human genetics studies. Whereas proponents of the Icelandic health-sector database stressed the homogeneity of the Icelandic population as a favourable asset for that project, the heterogeneity of the Estononian population was emphasized as an asset that would enable researchers to generate insights from the Estonian data that could apply to the rest of Europe (Dawson 2002; Fletcher 2004: 9). The initial aim of the project was to create a phenotype-genotype database of the Estonian population, including the data of 1,000,000 people. This goal was bold, as no suitable technology existed at the time for genotyping such a huge number of tissue samples at a reasonable cost. The project was scheduled to be carried out from 2000 to 2009, and during the first years it was planned to collect only the data and then wait for technological progress that would greatly decrease the genotyping costs.

The launch of EGP was in March 1999 with a project contract between the Estonian government and the foundation *Eesti Geenikeskus* (Estonian Genome Foundation), a nonprofit body founded in January 1999 by Estonian scientists, doctors, and politicians to support genetic research and biotechnology in Estonia and specifically to prepare the launch of the EGP. The initial proposal defined the following as the goals of the project:

- Reaching a new level in health care, reduction of costs, and more effective health care.
- Improving knowledge of individuals, genotype-based risk assessment and preventive medicine, and helping the next generation.
- Increasing competitiveness of Estonia – developing infrastructure, investments into high-technology, well-paid jobs, and science intensive products and services.
- Better management of health information databases (phenotype/genotype database).
- Support of economic development through improving gene technology that opens cooperation possibilities and creates synergy between different fields (e.g., gene technology, IT, agriculture, health care).

(Estonian Genome Foundation 1999)

Although 1999 was a perfect year for starting the project because deCODE Genetics had achieved worldwide attention and the international biotechnology industry was performing well, the specific strategy of the Estonian project remained rather vague and many management issues were still waiting to be solved.

It was also already acknowledged that the project could not be realized using the limited public resources of a small country like Estonia. Because of the favourable outlook of genomics research and the continual venture capital inflow into start-ups, the private–public partnership was regarded as the best solution. With this decision, a cornerstone of the Estonian Genome project, its funding had been left only to a limited extent to the unpredictable forces of the international biotechnology and venture capital markets. In return, the government promised to support the setup of a legal framework if the project could be financed by the private sector (Järvesoo 2000).

Setting up the biobank

The launching of the Estonian biobank was preceded by the creation of special legislation for the planned biobank, following the Icelandic path of securing genetic databank development by legislation. The so-called Human Genes Research Act (HGRA) was passed by the Estonian Parliament (Riigikogu) with forty-two yes votes and three no votes on 13 December 2000. The HGRA laid the foundation for the establishment of the Estonian Genome Project and was supplemented by several working documents regulating areas such as data handling and informed consent. The HGRA regulates the establishment and maintenance of Gene Bank, the organization of genetic research, protection of the voluntary nature of gene donation and the confidentiality of the identity of gene donors, and the protection of persons from misuse of genetic data and from discrimination based on interpretation of the structure of their DNA and the genetic risks arising.

According to the HGRA, genetic research relating to the Gene Bank is permitted to study and describe the links between genes, the physical and social environment, and the lifestyles of people; to find medicinal products or methods of treatment on the basis thereof; to assess individual health hazards; and to prevent illnesses (subsection 6 (1)). The use of the Gene Bank for other purposes, especially to collect evidence in civil or criminal proceedings or for surveillance, is prohibited (section 16). The act ensures that no one shall be discriminated against on the basis of genetic information (section 25–7). The law stipulates bringing up criminal charges for inducing a person to become a gene donor, carrying out illegal human research, disclosing secret information, and discrimination.

Furthermore, the act stated that gene donors shall decide whether they want to know their genetic data or not. The gene donor and a doctor treating the gene donor shall have the right to receive personalized information. This, it should be noted, is a provision that differs from many other biobank projects where strict anonymity is the rule. In its basic concept, the idea of a 'flow-back' of information from genetic analysis to patients is clearly considered. As one of our interview partners told us, the vast majority of the gene donors seem to have opted for this possibility during the early active phase of the EGP (Interview, 6 April 2005, General Practitioner, Tartu). The law furthermore regulates that a gene donor shall not receive any remuneration for participation in the project. Blood samples and health and genetic data are the property of the Gene Bank. To ensure confidentiality of a gene donor, the personal data of the gene donor shall be separated from genetic data, and each blood sample and set of health data shall be given a unique 16-digit code. If a gene donor does not want to participate in the Genome Project anymore, he or she shall have the right to demand deletion of the data that enable identification of his or her person or, in certain cases, of all the information stored in the Gene Bank about him or her. After deletion of the given data, it will not be possible to associate a blood sample and a gene donor, and the donor shall never receive any information about him or her.

The implementation of the project was started by the Estonian Genome Project Foundation (EGPF), a nonprofit organization founded by the Government of the Republic of Estonia in 2001 (Estonian Government 2001) under the provisions of a special legislation. The supervisory council of the EGPF consists of nine members, appointed by the Riigikogu, the Government of the Republic, and the Board of the Estonian Academy of Sciences.

The EGPF has also established the Scientific Advisory Board. The aim of the Scientific Advisory Board, established in October 2002, is to counsel the Supervisory Board and the Management Board in questions that contain scientific aspects. When necessary, the Scientific Advisory Board estimates the scientific validity of the scientific research carried out on the basis of the Gene Bank data.

According to the HGRA, a special Ethics Committee was established to assess the ethicality of the processing procedures of the Estonian Genome Project Foundation. The main aim of the Ethics Committee is to assist in ensuring the protection of the health, human dignity, identity, security of person, privacy, and other fundamental rights and freedoms of gene donors and resolution of general ethical problems related to human gene research.

Thus, by 2002 a coherent political governance structure for the Estonian Genome Project had been created in which the state played the key role of the regulator of the project.

Establishment of the EGPF

The difficult story of creating the financial basis for the EGPF must be seen in the context of a small, Post-Soviet, European economy with only very limited funding available for the research system. The government provided initial funding of 64,000 EUR for setting up the public foundation (EGPF) in 2001. The preparation and establishment of the biobank during 2001–2

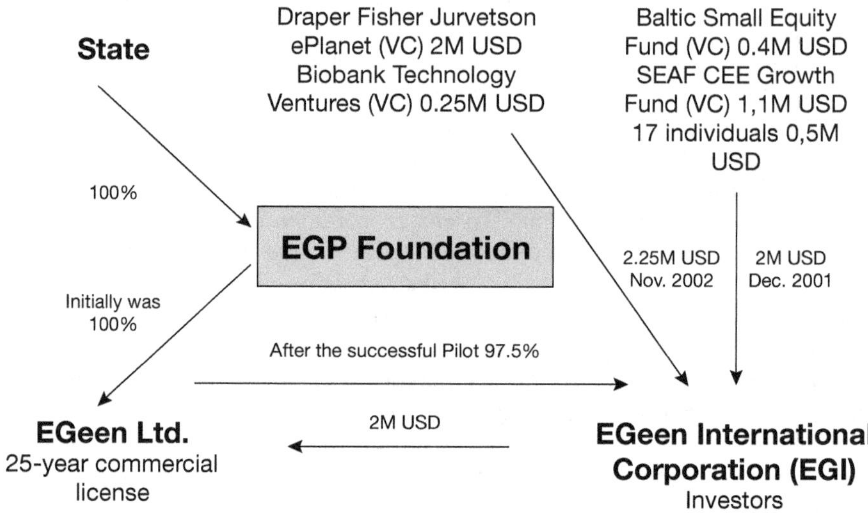

Figure 4.1 Initial financing and ownership scheme (before December 2004)

was funded by investors through the limited company EGeen, which was granted an exclusive twenty-five-year commercial license for using anonymous data of the biobank. Thc Estonian company EGeen was owned be the EGPF and EGeen International Corporation, located in California (USA).

During the preparations (2001–2), a laboratory and information system was built and GPs received intense training. Within the framework of the ISO quality management system, precise procedural instructions were worked out. The first tissue samples were taken from gene donors in October 2002. However, during the fourth quarter of 2003 the first conflicts in the consortium began to emerge. EGI (EGeen International Corporation) started to question the quality of collected data of about 9,000 gene donors and stressed the need to concentrate on specific disease groups such as hypertension. The conflicts between the EGPF and EGI continued during the next year, partially through the mass media. On 30 November 2004 the exclusive license and financing contract with EGeen and EGI was finally terminated, and the EGP was free of commercial strings, which made it possible to seek public financing (Estonian Genome Project Foundation 2004).

Data collection was actively carried out during six months of 2004 (February, March, and September to December). The planned target of 2,000 gene donors was not reached because of financial constraints that occurred during the spring of that year. In 2004 the Gene Bank received data of 1,501 voluntary gene donors. By the end of 2004 the biobank contained data of 10,317 gene donors. The anonymous data of gene donors started to be available for users already in 2004. There was no large scale data collection in 2005 and 2006. However, in February 2007 the large scale data collection process resumed and, in total, thirty-eight people were working in the reorganized organization. The EGPF offered services both to research institutions and companies of the pharmaceutical industry and participants in research projects. Cooperation projects have been set up with McGill University (Canada), the Paul Stradins Clinical University Hospital (Latvia), the Oncology Clinic of the University of Tartu Clinics (Estonia), the Women's Clinic at the University of Tartu Women's Clinic, and the Chair of Biotechnology at the University of Tartu.

Financing of the EGPF

Although the Estonian Genome Project began early to link the biobank project with public health by collecting data with the help of GPs and establishing a good connection between the project and society, its initial business model failed soon after the project was launched. The ensuing crisis that almost led to the termination of the project points to the fact that a viable business model is key for any biobank project. While the EGP was to spearhead the Estonian biotechnology industry, there was neither much of a national biotechnology industry to lead, nor any national venture capital industry prepared to invest in the EGP (Frank 2001). This created an overwhelming dependency on foreign venture capital to invest in a project whose commercial value was far from clear.

From 2001 to 2004 the cost of the project was about €4.6 million, which was mainly invested by private capital. During 2005 and 2006 the government provided an additional €0.84 million to store collected data and cover the costs of the EGPF. In total the project has used about €5.5 million, of which about 85 per cent has been provided by private capital (Estonian Genome Project Foundation 2004).

In late 2003, the first conflicts between EGPF and EGeen International Corporation materialized. EGeen started to question the quality of collected data of a portion of the gene donors and wanted to concentrate on research of specific disease groups, obvious targets for potential commercial application. It took about one year for the EGPF and EGeen to end its cooperation. In late 2004 the exclusive license and financing contract with EGeen was finally terminated, and in December 2004 the EGP was a public project seeking public funding and international and transnational collaboration opportunities (Estonian Genome Project Foundation 2004). At the end of 2004, because of unclear financial prospects, the collection of data was stopped, and the EGPF's management board had to lay off one-third of the foundation's employees. In addition, many of the employees were reduced to working part-time. No additional data was collected in 2005 and the first half of 2006. The Estonian Genome project seemed to have collapsed. In the beginning of 2005 the EGPF turned to the government for financing and presented a budget of €0.55 million to cover expenses for the year 2005. The annual budget that could allow active data collection was estimated to be about

€2.3 million. The maximum budget requested for the next three-year period was projected to be €8 million, which would enable further data collection and participation in collaborative research projects, allowing the sample volume of the EGP to be increased to 100,000 gene donors. Although the government had declared biomedicine to be a key policy area in its 2001 report *Knowledge-Based Estonia*, in January 2005 the finance minister gave a clear signal that the government was not interested in financing wide-scale data collection. However, the finance minister did promise to take care of the costs needed to maintain the project if necessary. In March 2005 the government decided to provide €0.27 million from its reserves to ensure the storage of collected samples. These steps kept the project from collapsing, but could not avoid a standstill of the development of the biobank.

While representatives of the government suggested in interviews carried out in April 2005 that the EGPF needed to redefine its working model and provide justification of the budget, including financial analysis of the costs and revenues involved (Lõuk 2005), not much help came from the mass media. The Estonian media defined the EGP (especially during the financing negotiations with EGeen) in their editorials as a business project. After the collapse of the private financing scheme, media and critics strongly stressed that the project was developed by Prof. Andres Metspalu and Dr Jaanus Pikani for their own benefit and in the interests of an American company. Therefore, the critics argued, the government should not provide resources for a project that had been abandoned by private capital (Äripäev 2004). The financial crisis of the project had translated into a deep crisis of public trust in the biobank project as such.

The negotiations about the future of the project were disrupted once again by the fall of the government in March 2005. The new government was faced with a project that it neither wanted to finance nor could shut down completely for political reasons. Private capital jumping in seemed to have been as unattractive as terminating one of the few high-visibility science projects in Estonia. This deadlock came to an end later in 2005 when the ministers of Social Affairs, Economic Affairs and Communications, and Education and Research declared in a combined press conference that the government must provide public funding for the project and formed a group of experts to design a detailed action plan (Ojakivi 2005). The fact that all three ministries were headed by one of the coalition parties facilitated the negotiation process within the coalition government with the other parties. The final decision to fund the project from the state budget was made by the government on 4 May 2006 (Estonian Government 2006). According to the decision of the government, the state will provide total funding of €7.67 million to continue large-scale data collection in the years 2007–10 to reach a 100,000 samples target by year 2010. The funding of the EGP is expressed in the budget strategy for the period 2007–13. The planned funding for 2007 in the draft state budget is about €1.2 million. The EGP has set a target of 30,000 samples by the end of 2007. For the year 2006, it provided €0.34 million for expenses and storing of available samples from the government reserves. Finally, in February 2007 the parliament passed the bill that enabled the reorganization of the public foundation into an institution of the University of Tartu. On 30 March 2007 Tartu University Eesti Geenivaramu was established. With guaranteed funding to reach 100,000 samples now, the remarkable percentage of about 10 per cent of Estonian population, it is planned, will participate in the EGP.

Collecting data for research and healthcare

An especially remarkable feature of the EGP in its full operation will be how it locates itself within the Estonian healthcare system. The way this is done represents a key feature of the Estonian strategy to govern its biobank by connecting research with healthcare applications. A specific aspect of the Estonian biobank project is how phenotype (health status and medical record data) and genotype data of voluntary 'gene donors' are collected into the gene bank. This tends to be a critical part in the governance of all biobank projects: how to connect patients with the biobank. As this book shows, different models exist. In the Estonian case (as in the UK case) a key role is played by the general practitioners (GPs) in linking the biobank project to the healthcare system, and thus the envisioned establishment of a novel system of preventive medicine. Accordingly, their enrolment into the project will be of utmost importance for the governance of the Estonian biobank.

The integration of the GPs into the EGP was one of the successes of the project in its early stage before data collection had to be aborted because of the financial crisis. A central motive for their participation in the project seemed to be the prospect of the GPs being part of a prestigious medical–scientific project. Furthermore, the GPs received incentive to participate, as all GPs received a €32–34 financial compensation per donor. Especially interesting is the fact that if a gene donor agrees, the GP as data collector would receive access to the genetic data of his or her patient. In order to participate in the biobank project, the GP must obtain permission from the Data Protection Inspection to process the personal data (Interview, GP, 6 April 2005). However, these aspects of the EGP never materialized because, due to the funding problems, only the data collection proceeded, not linkage analysis or any flow-back of information to the GPs. It remains to be seen how these aspects of the EGP will unfold after the resumption of the project in 2007. By July 2007 the EGP had signed contracts with 434 data collectors of the 1,021 total number of GPs.

The integration of patients into the project operated in the following way, all clearly regulated by law: a person who is interested in participating in the EGP must contact his or her GP and make an appointment, or people are asked by their family physicians to become gene donors. Upon visiting the GP, the person is first introduced to the EGP. After deciding to participate,

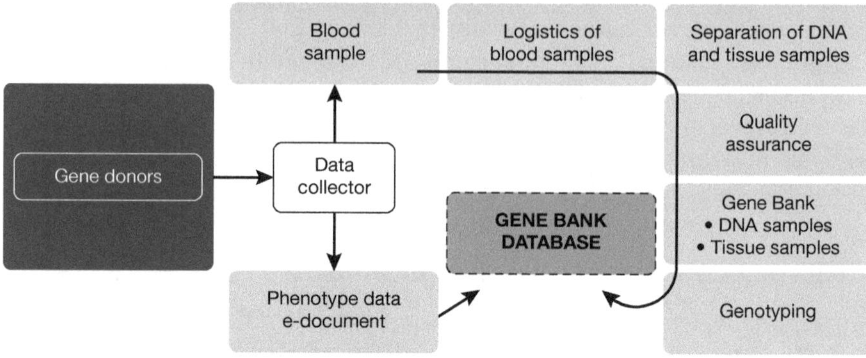

Figure 4.2 Data collection process

they must sign the informed consent form, respond to the computer-assisted questionnaire, and then donate a 50-ml blood sample.

The data collection process at the data collector's office consists of three steps:

1 Obtaining a Gene Donor Informed Consent.
2 Taking a tissue sample.
3 Filling in the questionnaire of Health and Genealogical Data.

At the first step the doctor introduces the EGP and explains its goals to the person who wishes to become a gene donor. The doctor then discusses the role of a gene donor in the project. If the person wishes to become a gene donor, he or she signs the Gene Donor Consent Form.

After the signature, the data collector fills in the digital Questionnaire of Health and Genealogical Data by interviewing the gene donor. The questionnaire consists of seventeen modules and contains in total 182 questions, and the question order is created by the Computer Aided Personalized Interview. Completion of the questionnaire takes approximately 1–1.5 hours. Health questions are in accordance with the ICD10 structure. The completed questionnaires are sent to EGP via the Internet in the form of encrypted documents. Personal data from questionnaires delivered to the Gene Bank are separated and replaced with a 16-digit code[1] in the coding centre. Health data that has been separated from personal data are stored in the database of the Gene Bank. A tissue sample (50 ml of blood) is taken from a gene donor before or after completion of the questionnaire. Each tissue sample is marked with a special bar code. Samples are stored in the data collector's office in a definite temperature range (6 ± 2 °C) until its delivery to a transportation company. A courier service delivers tissue samples to EGP within thirty-six hours from the moment of taking the tissue sample, observing the security and other regulations established. The total amount of DNA separated from each tissue sample is measured, and the quality testing of the DNA is carried out with three different methods. The results obtained are stored in the EGP database in a digital form, which allows the EGP to supply each DNA sample issued by the gene bank with a document proving its quality. The data are stored in the database under the storage addresses. The data of gene donors are stored anonymously in the EGP database and they can be related to the gene donor, that is decoded, only in cases stipulated in the Human Genes Research Act. Following these protocols, the EGP is positioned in a link between the Estonian healthcare system and the research system.

Interacting with society

It is interesting to note that the Estonian biobank project has made a deliberate and structured effort to communicate with 'the Estonian public', which mainly took the form of information dissemination from the EGP to the public, supported by monitoring through public opinion polls. However, it was by no means clear or obvious that this project would attempt to reflect the social context it is embedded in.

To introduce the idea of the biobank project to the public, several specific television and radio programmes and series of articles were initiated to reach the population. Experts explained the basics of genetics and the international developments of the field. Estonian media first covered the topic of the EGP in spring 1999. The first national daily newspapers to write about it were *Postimees* and *Eesti Päevaleht*. By autumn 1999, the topic of the EGP had reached most news channels, and the first opinions were published. Since that time, the Estonian media have published a number of articles and opinions that have been directly or indirectly related to the EGP, and the topic has also been covered in radio and television programmes. Critics of the EGP have blamed the promoters of the project in their opinion articles for creating propaganda and advancing a professional public relations strategy (Tasmuth 1999). At the time of launching the EGP, during 2000–1, the Estonian Genome Foundation undertook several initiatives to raise the profile of the project in the media. The projects included newsletters, television, and radio programmes. The 'Geenileht' (Gene Paper) newsletter was published four times as part of *Loodus* (Nature) magazine (circulation about 2,500) and twice as an insert to the newspaper *Postimees* (circulation about 65,000) in May 2001 and August 2001. In addition, a special genetics-related programme series was broadcast on the Estonian National Television. A five-part *Geenistuudio* (English translation Gene Studio) was shown from September to October and a ten-part *Telegeen* (English translation Telegene) from October to December in 2000; the re-runs were shown in spring 2001. In October 2001, Estonian National Television showed three parts of the *Kuumad Geenid* (English translation Hot Genes) series and the sequels in spring 2002.

Further, a number of public opinion polls were also conducted on the biobank project, for example, University of Tartu commissioned an opinion poll in December 2002 in the context of a fifth EU Framework Programme project.[2] These polls showed a majority of the population being supportive of the project, but also large parts of the population being undecided or feeling uninformed. Overall, interacting with society had clearly been an operative goal of the Estonian Genome Project but had taken mainly the form of information dissemination.

Conclusion

Although not as much in the limelight as the Icelandic health database project, the Estonian biobank project nevertheless deserves attention. It demonstrates the complex challenges for biobank governance that go far beyond the establishment of a coherent and well-functioning legal-ethical framework. Obviously, establishing such a framework was a rather smooth process in Estonia, where the parliament passed special biobank legislation with much political diligence. The proponents of the biobank project had managed successfully to put the EGP into the core innovation policy agenda of Estonia by presenting the EGP as a biotechnology breakthrough with the potential of contributing significantly to economic growth and development. The EGP was framed as not only key for Estonia's future economic development, but also defined as 'a bridge to Europe' and, thus, as a way to overcome the traumatic political experiences of the past. At the same time, the Estonian biobank project also established a particular model of linking patients with the biobank through general practitioners thus allowing a flow-back of patient genetic information to the general practitioners.

However, the project almost collapsed because of a business model in which most of its financial resources were to come from foreign venture capital. When in 2000 the international biotechnology bubble burst and the biotechnology industry underwent a period of crisis, this prospect virtually vanished. This constellation put the Estonian government in a no-win situation. On the one hand, the government was not prepared to spend large parts of the nationally available science and technology funding on the EGP, but on the other hand the government did not want to oversee the collapse and thus failure of the project, which would have been disastrous for Estonia's internationally desired reputation as a hub of innovation. In other words, the government had become hostage to the EGP, not an uncommon experience with large-scale scientific-technological projects.

Precisely because the Estonian Genome Project had been presented to the Estonian citizens and the international scientific community as a genuine achievement and demonstration of 'Estonia's return to Europe', and because a sizeable sample had already been established, to abandon the project for financial reasons was not a tenable option. The Estonian case underlines the point that depending on the type of research funding system and given financial conditions, a viable business model needs to be a key element in any biobank governance structure.

Notes

1 Coding means the replacement with a unique code of such data (name and personal identification code), which enables the direct identification of the person. The code has sixteen digits, and it is created by a computer by using incidental numbers and letters. The code is adhered to the Consent Form. Coding is one of the security measures for protection of the data.

2 The project aim was to consult citizens by providing knowledge of public views of privacy and related moral values in the context of human genetic databases. See www.ut.ee/eetikakeskus/?eng/tegevus/21#ELSAGEN

References

Aalto, P. (2001) 'Post-Soviet geopolitics and the politics of identity in democratic consolidation: the case of Estonia', in D. Berg-Schlosser and R. Vetik (eds), *Perspectives on Democratic Consolidation in Central and Eastern Europe* (pp. 107–16). New York: Columbia University Press.

Äripäev, 'Geenivaramu kolm tilka verd sai, aitab talle', *Äripäev,* 6 October 2004.

Dawson, E., Abecasis, G. R., Bumpstead, S., Chen, Y., Hunt, S. *et al.* (2002) 'LD across human chromosome 22', *Nature*, 418: 544–8.

Estonian Institute, 'Web encyclopedia Estonica', available online at www.estonica.org/ (accessed 22 October 2006).

Estonian Genome Foundation (1999) 'Letter to the Estonian Government from Estonian Genome Foundation', 9 September.

Estonian Genome Foundation. Website: www.genomics.ee (accessed 22 October 2006).

Estonian Genome Project Foundation (2003) 'Eesti Geenivaramu osaleb maailma mõjukamate geeniprojektide katusorganisatsiooni loomisel' (17 July), available online www.geenivaramu.ee/index.php?show=uudised&sub=arhiiv&id=120&lang =est (accessed 22 October 2006).

Estonian Genome Project Foundation (2004) 'Geenivaramu lõpetas koostöö senise rahastajaga' (24 December), available online www.geenivaramu.ee/index.php? show=uudised&sub=arhiiv&id=170&lang=est (accessed 22 October 2006).

Estonian Genome Project Foundation, Website: www.geenivaramu.ee (accessed 22 October 2006).

Estonian Government (2001) 'Order nr 177-k 13 March 2001', published in *State Gazette* (20 March 2001), available online at www.riigiteataja.ee/ert/act.jsp?id= 84864&replstring=33 (accessed 22 October 2006).

Estonian Government (2006) 'Minutes of 4 May 2006 meeting'.

Estonian Research and Development Council (2006) 'Minutes of 22 February 2006 meeting', available online at www.riigikantselei.ee/failid/OTSUSED.doc (accessed 30 May 2006).

Fletcher, A. (2004) 'Field of genes: the politics of science and identity in the Estonian genome project', *New Genetics and Society*, 23 (1) (April): 1–12.

Frank, L. (2001) 'Biotechnology in the Baltic', *Nature Biotechnology*, 19: 513–15.

Järvesoo, V. (2000) 'Geenivaramu asub otsustavalt doonorite arvu suurendama', *Äripäev* (12 June).

Lõuk, K. (2005) 'Riik ja geenivaramu – kuidas edasi?', *Eesti Päevaleht* (8 February).

Meri, L. (1999) 'What does Estonia produce?', *Speeches of the President of the Republic, 1992–2001*, available online at http://vp1992–2001.vpk.ee/eng/k6ned/K6ne.asp?ID=4295 (accessed 30 May 2006).

Ojakivi, M. (2005) 'Ühe partei soolo Geenivaramu rahastamisel tekitab segadust', *Eesti Päevaleht* (8 December).

Tasmuth, T. (1999) 'Eesti geenivaramu loomine teeb ettevaatlikuks', *Eesti Päevaleht* (18 October).

Tiger Leap Foundation, Website: www.tiigrihype.ee (accessed 22 October 2006).

Unwin, T. (1999) 'Place, territory, and national identity in Estonia', in H. Herb Guntram and David H. Kaplan (eds), *Nested Identities: Nationalism, Territory, and Scale* (pp. 151–74). Lanham, MD, and Oxford: Rowman & Littlefield Publishers.

5 Patient organizations as the (un)usual suspects

The biobanking activities of the Association Française contre les Myopathies and its Généthon DNA and Cell Bank

Michaela Mayrhofer

Introduction

The general aim of this chapter is to address the biobanking activities of the French patient organization *Association Française contre les Myopathies* (AFM). It seeks to put these activities into the broader context of biobanking activities in France. In particular, this chapter is concerned with the discourses and practices[1] of the *Généthon DNA and Cell Bank* and how meaning is given to these. Ultimately, this chapter is about how both governance of biobanks and governance through biobanks[2] is exercised. Methodologically, it draws mainly on primary documents, and observations and interviews that I have conducted with key actors in the field.[3]

To begin, what is a 'biobank'? Anne Cambon-Thomsen depicts biobanking as the 'organised collection of biological samples and associated data' (Cambon-Thomsen 2004: 866), which signifies collecting, registrating, conserving, storing and utilizing biological material in an organized manner. A biobank is thus an institution that exercises the activity of biobanking. Worldwide, biobanks have become a major issue in the field of life sciences and countries with large-scale population biobanks have attracted most of the public and academic attention, particularly in relation to issues of informed consent, benefit sharing and data protection (Knoppers 2003; Godard *et al.* 2003; Williams and Schroeder 2004). However, these large-scale population biobanks represent only one category of a broad variety of worldwide biobank initiatives (Hirtzlin *et al.* 2003). Biobanks differ in scale, purpose (research, therapeutic or diagnostic), regulatory framework (explicit law, a set of regulations or best practice protocols), mission goals (gene mapping, drug development, service institution for research teams), material (blood, cancer tissue, tissue slides, bacterial strains, therapeutic substances or DNA), and stakeholders (the pharmaceutical industry, hospitals, research groups or patient organizations). Furthermore, biobank initiatives differ both in the modes of governance and the degree of involvement of the various stakeholders that finance, operate, and/or govern a biobank. Additionally, those stakeholders rely on different discourses and practices to justify their involvement in biobanking.

Generally speaking, biobanking can be described as a discursive practice that is both vague and open to articulation because each particular stakeholder embedded in a particular context articulates its biobanking activities differently. Private cord blood-banks, for example, embedded in the context of new biomedical therapies, articulate the practice of cord-blood banking as a kind of biological investment or 'venture capital', offering parents the opportunity to bank their child's cord-blood for potential later use (Waldby 2006). Consequently, '[s]amples of tissue and patient records are far more than simply abstract "bytes" of data' according to Nik Brown and Andrew Webster, who suggest that '[r]ather, they are highly corporeal, even visceral, substances loaded with meaning and situated in a nexus of value that can be changeable over time' (Brown and Webster 2004: 95). By pointing out that samples are situated in a 'nexus of value', Brown and Webster refer to the contingent character of the meaning of collected samples, which is not only changeable over time but also across contexts: the value of collected body material might be transformed over time, change drastically with the emergence of new technological tools, differ in a different institutional setting, and become meaningful in ways yet unknown.

Historically, pathology departments were concerned with diagnoses of patients' specimens. The introduction of molecular biological techniques has opened up a new dimension in the retrieval of information hidden in the preserved sample material (Sundström 2001), information that has become useful to other users, such as the pharmaceutical industry. As a consequence, collected specimens in pathology institutes have obtained new meaning and value. The *Pathologieinstitut Graz*, for instance, has recently rearticulated its historical collection of human tissue as a 'key resource for genome research' (Medizinische Universität Graz 2006). Another example of the rearticulation of meaning accompanying medical progress is that of umbilical cord-blood in the clinical context. Historically, it 'was once considered an abject tissue, designating neither mother nor child, it is now deemed a significant fragment of the infant' (Waldby and Mitchell 2006: 110). As a consequence, its meaning and value is rearticulated and thus transforms cord-blood, which was formerly depicted as a clinical waste product, into a valuable substance for the treatment of serious blood disorders. Catherine Waldby (2002) conceives the value of biological material reformulated through the biotechnological modification of the potential biological capacities of body fragments as 'biovalue'. Among other things, 'biovalue' captures the potentiality of the collected material. It does so by referring to the new, yet unanticipated, possibilities that future biotechnological inventions might provide and thus offer a conceptual tool to understand the contingent character of biobank activities. Thus, 'biobank' can be understood as an empty term in the biomedical discourse whose meaning is contingent and context dependent. In other words, the concrete meaning of biobanking is discursively constructed within the meaningful web of a biobank's particular discourse and practices. Through articulation, actors impose concrete meanings onto practices, objects or signifiers that do not hold per se any particular meaning in themselves (Laclau and Mouffe [1985] 2001). Therefore, the practice of biobanking should be surveyed against the background of the biobanking institution and the particular context in which it is embedded.

In the case of the *Généthon DNA and Cell Bank*, the biobank's activities are articulated by the practices and discourse of the patient organization AFM and are embedded in the context of biomedical research. In general, the degree of involvement of patient organizations in biobanking activities can range from providing a gateway for research teams by recruiting donors to governing a biobank entirely. The US-based *Angioma Alliance*, for instance, has been involved in facilitating blood or tissue donations to research laboratories by providing information on research projects and encouraging its clientele to donate (see Fletcher, Chapter 7). The patient organization *International Myeloma Foundation* (IMF) has moved a step further by setting up its own repository of human material in the form of the *Bank On A Cure*. Yet, what motivates patient advocacy groups getting involved in biobanking activities and biomedical research activities in the first place? Alastair Kent provides the following insight:

> Patient support groups and the individuals and families who belong to them are amongst the keenest advocates for biomedical research. It is not difficult to see why. Finding that you, your child or another close member of your family has a genetic disorder automatically makes you a member of a 'club' that you did not ask to join, that you are desperate to leave, but, as things stand at present, from which you cannot escape.
> (Kent 2002: 707)

Hence, in this situation, the hope to escape becomes a powerful driver of participation, and allows patient organizations to operate within a 'political economy of hope' (Rose and Novas 2005: 454). People are mobilized in various ways relying on the promise of biomedical research to provide a better future, a cure. Sometimes, this programmatic claim is expressed in the naming of a patient organization's biobank, such as the *Patients Tumorbank of Hope* (PATH), initiated by the German patient organization Mamazone. The naming contributes to the articulation of the meaning given to biomedical research and biobanking activities. It signifies the conviction that a cure can be found through biobanking. According to Nicolas Rose and Carlos Novas, this 'political economy of hope' is the manifestation of a new kind of citizenship, which they call 'biological citizenship', that links the 'conceptions of citizens to beliefs about the biological existence of human beings as individuals, as families and lineages, as communities, as population and races, and as a species' (Rose and Novas 2005: 440). The biological aspect of their being becomes the basis for a shared conception of identity (Rose and Novas 2005: 440).

In other words, one's biological future – one's disposition to disease – although possibly genetically determined, is no longer understood as a matter of fate. In this spirit, I want to compare biobanks to nation states insofar as they provide a space, a territory, for Novas and Roses' biological citizens. I consider this analogy useful to the extent that it exemplifies biobanks as locations of the 'political economy of hope' promising a better future through biomedical research. Continuing with the analogy of the nation state, the disease is disciplined as the 'enemy' or 'the other'. As is typical of patients with rare diseases, 'the other' is not somewhere out there, but within the very body of each 'citizen'. Some authors have argued that 'the other' can be within the body of the state, yet, in this particular case the enemy is literally within the individual's body and therefore creates new groups ('biological citizens') and categories of people.[4]

Furthermore, it has been emphasized by several authors that, in the era of genetics and genomics, health has become a moral obligation, articulated as the moral responsibility of the individual to remain healthy (Crawford 1985), to manage one's fate (Strauss *et al.* 1984), rather than to recover from illness (Parsons 1951). In this way, health becomes a matter of ongoing self-transformation and self-governance (Clarke *et al.* 2003: 172). This transformation has enormous impact on the meaning given to research endeavours, and on the analysis of biobanking activities of patient organizations. Indeed, from this perspective activities of patient organizations should be taken as manifestations of 'biological citizenship' and as locations of self-governing practices (governing the body through biobanks, mode 2, see Gottweis and Petersen, Chapter 1).

In recent years, the participation of patient organizations in biomedical research has become more active, insofar that certain patient advocacy groups are involving themselves directly in strategic decisions concerning research. In France, the AFM is a notable patient organization whose engagement in research is depicted as a 'partnership model' of direct civic participation with science (Rabeharisoa 2003). Firstly, the new model emphasizes patients as equal partners of experts and scientists in knowledge generation. Secondly, it identifies patients as decision-makers of their own research policy, which 'does not delegate to them [the scientists] decision-making powers on the definition, management and evaluation of its research policy' (Rabeharisoa 2003: 2132), but instead claims the decision-making power for itself at all times.[5] These features thus position the AFM as a very particular case of a patient organization's involvement in biomedical research.[6]

In what follows, I will give firstly an introduction to the French biobanking 'industry' and its overall characteristics in order to position the activities of the AFM therein. In doing so, I aim to answer the question of whether or not the involvement of the AFM is also something particularly French or something that could emerge against the culture of French etatism.[7] In a second section, I examine how the AFM is involved in biobanking activities. This section is guided by the following questions: Why does the AFM see biobanks as beneficial for its cause? In what ways does it engage in biobanking and how does it articulate its actions? What modes of governance are employed? What discursive practices guide the daily routine of biobanking? To answer these questions, a special focus will be put on one particular biobank, namely the *Généthon DNA and Cell Bank*, because it promises interesting insights in governance matters. Furthermore, it represents both an articulation of the 'partnership-model' and the concept of 'biological citizenship' as this chapter will show.

Section 1: the French context, biobanks for research and the particularity of the AFM

In the introduction I argued that, among other things, all biobanking activities have to be evaluated within their specific context. This means that the *Généthon DNA and Cell Bank* has to be seen, first of all, as embedded in the overall context of biobank activities for research purposes in France. Hence, I will elaborate on biobanking in France on a general level, showing the characteristics of the French biobanking 'industry' and the categories of biobanks that exist in France. Focusing on the AFM, I will show that it is a quite special stakeholder in the context of biomedical research.

In reviewing France's involvement in biobanking on a general level (Hirtzlin *et al.* 1999; Moutel *et al.* 2000; de Montgolfier 2002; CCNE 2003; Hirtzlin *et al.* 2003, Pontille *et al.* 2006), there seems little to suggest anything particular compared to other countries, such as Germany (Nationaler Ethikrat 2004, see Schneider, Chapter 6). France has neither planned nor is planning a large-scale population biobank, like the UK has (Petersen 2005). Like most countries worldwide, it has a broad variety of biobanking initiatives, ranging from small collections in public hospitals to large private non-profit biobanks, widely differing in scale and purpose (e.g. research, therapeutic, diagnostic), relying on a web of regulations[8], their governance broadly based on Best Practice Protocols.

What about public debate on biobanking issues? Although the *National Consultative Ethics Committee for Health and Life Sciences* (CCNE) has produced an opinion on biobanks (CCNE 2003), public debate on biobanking is, as in most countries, basically nonexistent in France. An explanation for this lack of public interest might be that the current French policy aims at a re-evaluation and re-structuration of already existing collections of biological material within the clinical and research context rather than creating a new large-scale biobank in an entrepreneurial manner. Historically, research centres such as the *Institute Pasteur* (Latour 1988) or the *Centre d'Étude du Polymorphisme Humaine* (CEPH) place France in a prominent position in the early days of organised biobanking in relation to research activities. In 1993, the CEPH's collection of human genetic material established the basis for the first physical map of the human genome (Cohen 1993).[9] Today, the French government is highly active in this field at the international level, leading the OECD *Task Force on Biological Resource Centres*. In accordance with the OECD initiative, it has launched national 'cohortes et collections' support programmes encouraging already existing biobanks to evolve into *Biological Resource Centres* (BRCs). These programmes had a total budget of €5.3 million and were directed by the Ministry of Research and the *Institut national de la santé et de la recherche médicale* (Inserm).[10]

So, what are the characteristics of biobanks in France? The French biobanking 'industry' is characterized by three main actors: (1) biomedical research companies and the pharmaceutical industry (e.g. *Aventis, Genset Serono*), (2) public healthcare and research institutions (e.g. public hospitals, Inserm) and (3) private non-profit institutions (e.g. patient organizations, such as the AFM). Often, these actors are interconnected, and show a variety of modes of governance. A stakeholder that is the operator in one context, might act as the initiator in another. The *Banque Cassini* at the Parisian *Cochin Hospital,* for instance, has been sponsored and initiated in its current form by the AFM. Today, however, it is governed and financed exclusively by the *Cochin Hospital.* The AFM has no influence on its modes of governance. Things are different in the case of the *Généthon DNA and Cell Bank.* Here, the AFM is both initiator and main sponsor at the same time. Also, some biomedical research companies and the pharmaceutical industry operate their own biobanks, while others (additionally) obtain access to samples at biobanks such as the *Biobanque de Picardie.*

A study by Grégoire Moutel and others (2000) concluded that more than twenty departments at both the *Necker University Hospital* and the *Reims Hospito University* possess a DNA collection.[11] Isabelle Hirtzlin and others (1999) observed an expansion of human biological sample collections since the mid 1980s, which they argue is explained by the emergence of molecular biology and genetics. In their survey, they identified forty-one Inserm and *Assistance Publique-Hôpitaux de Paris* (AP-HP) laboratories that store samples 'usually [. . .] as a side activity to their clinical or research activities' (Hirtzlin *et al.* 1999: 3). Additionally, biobank operators are often unaware of their colleagues' activities in this respect, sometimes even in the same institutional setting, as the following example given by a French biobank administrator illustrates:

> Well, in this hospital complex I know of 3, no 4 biolibraries[12] . . . but there could be more. All you need is a fridge, isn't it? . . . And over the years stuff is gathered, you know . . . and then genetics made a lot . . . well, transformed as one can say a personal hobby into a biolibrary, you see?!

In other words, biological material was gathered in a rather unsystematic manner 'often as a side activity without a designated budget' (Cambon-

Thomsen 2003: 25). Traditionally, samples and data have always been collected as part of the ordinary practice of medicine: 'financed as part of the medical act by the health insurance system. The data is processed or communicated to other medical teams in a general climate of mutual trust.' (CCNE 2003: 9). Thus, genetics and recent technological developments led to a shift in the value perception of the stored biological material and enabled the production of 'biovalue'. Overall, recent developments in relation to the BRC initiative indicate that French policies seem to focus on research biobanks (as opposed to therapeutic or diagnostic biobanks), as preliminary research lead by Isabelle Hirtzlin (Hirtzlin *et al.* 1999), Grégoire Moutel (Moutel *et al.* 2000) or David Pontille (Pontille *et al.* 2006) has indicated.

According to French researchers, arguing in compliance with the French etatist culture, biomedical and genetic research is neglected by public funding. Although French governments have promised repeatedly to increase the overall research budget, which seems to align with heavy investment in the biotechnology sector, researchers remain doubtful about the effects of such announcements. So far, France has been ranked third in the European biotechnology sector with about 270 biotech companies of which almost 80 per cent are related to biomedicine. But, with little attention given to rare disease patients (Rabinow 1999), the patient organization AFM, advocate for neuromuscular disease patients, followed a do-it-yourself approach.[13] In the words of an AFM employee, the AFM decided in the late 1980s to act itself[14] and 'started to finance and work with researchers to advance genetic research, which was not very well supported by public research and not really by the government'. This statement contains three interesting themes: first, the theme of criticism of the state's research policy (which can almost be called a tradition among French researchers); second, the theme of the AFM as a self-empowered decision maker, which decided to pursue its own research policy; and third, the theme of equal partnership between the patient organization and the researchers in compliance with the 'partnership model'. Also, all three themes point out the unusual nature of this patient organization, which is particular in the French context; namely, it does not follow the conventional research way in the French context, which is to advocate state support. Consequently, if the question of whether the AFM is a typically French institution or patient organization, the answer would be simple: it is not.

Section 2: the AFM, genetic and rare disease research and biobanks

In the first section of this chapter I have examined the French context and concluded that it is not too different from other countries in relation to biobanks. Also, I have argued that biobanking is an activity pursued by many stakeholders in various contexts and that the majority of them can be summed up in the category of research biobanks. Among the stakeholders, the patient organization AFM seems the most active one in the context of biomedical and genetic research. In this section I will show how the AFM justifies its biobanking and research activities. Finally, I will examine how meaning is given to the biobanking practices of the *Généthon DNA and Cell Bank*.

The AFM, in concluding that the state would not act or was very inadequate in supporting rare disease research (AFM, DNA collection, Online. Available www.afm-france.org, accessed 20 August 2005), strategically decided to participate in genetic research. Historically, the AFM decided to become actively involved in research policy and genetic research in the mid 1980s (Rabeharisoa and Callon 1999), which it sees as the pathway to cures. Although the AFM's devotion to genetic research is articulated for the benefit of its rare disease clientele, its research policy is not limited to this alone. The logic behind its policy is that all kinds of genetic research will prove useful for rare disease patients and therefore need to be fostered.

> Since neuromuscular diseases have [a] genetic origin, AFM takes part in the development of scientific tools for the study of genetic and rare diseases in general, and provides scientists with new means and knowledge to accelerate our understanding of genetic diseases and open up treatment avenues based on an understanding of the genes.
>
> (AFM, Our Mission, www.afm-france.org, accessed 27 September 2006)

The particularity of the AFM's research policy lies in the rather unique decision to foster all kinds of genetic research, regardless of whether it has direct benefit for its clientele. Usually, patient organizations tend to support disease specific research (for example, breast cancer research, such as the German patient organization *Mamazone*), whereas the AFM fosters research on genetic diseases in general.

Believing in the power of genetics and genetic research, biobanking is depicted as yet another means to fight disease and as empowering practice. So, the AFM declares that '[s]ince its creation, the AFM is set on pursuing a challenging goal: curing neuromuscular diseases. It is motivated by the conviction that a cure is possible' (AFM 2004: 3). Thus, the AFM articulates genetic research as a key activity to find a cure. A fundamental and foundational value of the patient organization is to work 'against oblivion and ignorance' expressed in the 'parents' *refusal* to give up and to *accept* these diseases as their *destiny*'. It proclaims that the 'AFM's values are shared by parents and patients who are determined to take *all possible steps to combat* neuromuscular diseases' (AFM, Our Mission, Online. Available www.afm-france.org, accessed 27 September 2006, emphasis added). In the discourse of the AFM the disease is disciplined as the 'enemy' and by employing a warfare-like rhetoric (Rabeharisoa and Callon 1999), it positions the patients as active citizens or 'warriors' in both compliance with the 'partnership model' and the concept of 'biological citizenship'. The goal, to find a cure, is thus to be achieved through self-governing practices, which should liberate the patient from their biological determination, as articulated in the following:

> The AFM is born from a new generation of patients and patients' relatives who have decided to *take their destiny in hand* and to put up a resistance to the disease on all fronts. Engaged in scientific research as well as in assistance to patients, it acts independently and is guided solely by the interest of the patient.
>
> (AFM 2004: 3, emphasis added by author)

Following the same logic, the AFM concluded that biomedical research lacked genetic material in both quality and quantity and so started its biobanking activities in 1990. Since then, an AFM employee explains, it has organized sampling campaigns and started several DNA banks for the storage of human genetic material.

The AFM and the Généthon DNA and Cell Bank

We saw in the previous section that the AFM articulates its involvement in biomedical and genetic research as another step to find a cure for rare disease patients in compliance with the 'partnership-model' and 'biological citizenship'. In this context, biobanking is understood as a basic condition for research and consequently as a means to make it possible to find a cure. In this section, I will elaborate on the biobanking activities of the AFM in general and will show how meaning is given to the particular case of the *Généthon DNA and Cell Bank*, which, I argue, is an exemplification of the 'partnership-model' approach and the concept of 'biological citizenship'. I will argue that the meaning of the *Généthon* bank cannot be understood independently of its governing institution. Ultimately, I will examine the *Généthon DNA and Cell Bank's* modes of governance, governance of and governance through biobanks.

By its own account, the AFM has financed more than 600 DNA collection programs and initiated ten DNA banks inside France (Barataud 1999: 43), and has supported the creation of biobanks across Europe and French overseas departments and territories (AFM, DNA collection. Online. Available www.afm-france.org, accessed 20 August 2005). The most notable is perhaps the *Généthon DNA and Cell Bank* that is the only biobank over which the AFM has retained complete control. Therewith, the biobank exemplifies an articulation of aspects of the 'partnership model' as decision-making powers are retained by the patient organization. Also, the biobank is articulated as a site of 'biological citizenship' as it forms a new group around a biological conception of citizenship.

Institutionally, the *Généthon DNA and Cell Bank* is part of *Généthon*[15], a private non-profit organization that is almost entirely funded by the AFM.

The biobank is located in the *Généthon* building in Evry, in the same complex as the AFM itself, just a few kilometres outside Paris. *Généthon* is also an element of the biotech cluster *Genopole*[16] located at the same site. Notably, *Généthon*'s is chaired by Bernard Barataud, former president of the AFM, who is still a member of its *Administrative Council*. Clearly, these aspects of *Généthon* demonstrate inseparable ties between the AFM and *Généthon*, and the particular context in which the *Généthon* biobank is embedded.

Pursuing its objective to find a cure, the AFM relies on its enormous financial resources gathered through *Téléthon*, a yearly fundraising festival on French television. Thus, *Téléthon* is not only a money-making machine for the patient organization. Alain Kaufmann argues that both scientists and the general public make evident the support for the AFM and *Généthon* by their commitment to *Téléthon*:

> [This organization] transformed simultaneously the dynamics of research groups involved in localizing genes and the citizen's awareness of genomics expressing their generosity by giving money to the AFM. This can be interpreted as a renewal of the Pasteurian tradition of science as a public good.
>
> (Kaufmann 2004: 147)

What Kaufmann says in relation to commitment of both scientists and the general public is equally true for the staff of the *Généthon DNA and Cell Bank*. For instance, the commitment to *Téléthon* is visible at the location of the biobank, where a visitor can see on each wall posters of recent and previous *Téléthons*. This also contributes to a corporate feeling, which is articulated by a biobank staff member as follows:

> We are not a research group, we are sponsored by a patient organization, and we are at the other side of the wing, where all the handicapped and ill children are, and we meet in the cafeteria, we eat together, we are all together, and we all know that we are here to help these sick persons. Our director is a father of a sick child. [. . .] It is truly an associative ambiance, and we all work together to advance research and to help the children. Each time a child dies it is truly a trauma, we all know one another, that is really hard then, but it keeps us doing our work. . . Also for the Téléthon, we stick together, we are all there, the adults and the little ones, everybody, for the goal that it advances, that we gather money to continue.

It is striking that in the above statement by a staff member of the *Généthon DNA and Cell Bank* it is hard to distinguish to whom precisely the staff member is referring when talking of 'we' or 'us'. Whereas at the beginning, the interviewee clearly talks about the biobank, later on it is not clear whether the interviewee is referring to the biobank, the patient organization or to

Généthon. But, why is this important? It underlines the fact that the activities of the biobank cannot be understood as detached from the patient organization. Thus, I take this statement as an articulation that the identity of the patient organization and all it stands for is also apparent in the biobank. Staff members perceive their work as meaningful and articulate it in the spirit of the patient organization's goal to find a cure.[17] Also, the above statement exemplifies that biobanking must be understood as both context dependent and contingent.

Looking at this particular context reveals that the *Généthon DNA and Cell Bank* is not only unusual for a biobank, it is also unusual for a patient organization's biobank. Usually, patient organizations tend to delegate decision-making powers to a scientific board or locate the biobank out of their reach, as does the *Bank On A Cure* of the *International Myeloma Foundation* (IMF).[18] The *Généthon DNA and Cell Bank*, however, does not delegate its decision-making powers to experts. In its daily routine, according to one of its staff members, the *Généthon DNA and Cell Bank* collects, prepares, stocks, and distributes genetic material from patients with genetic disorders and their family members:

> We immortalize the cell lines, to have them available . . . to ask for each research project each time for a new donation would pose a logistical problem. They [donors/sick persons and/or their family members] change their address . . . Unfortunately, they have genetic diseases that evolve very fast and are very severe, sometimes the person already died, they are no longer there, they disappeared. On the other hand, research, which is a long-term research, genetic research, research on genes and all that, to study all the functions of a gene . . . We decided to immortalize the cells [with the Epstein-Barr virus] of our sick persons, and that allows us to reproduce indefinitely.

Samples are provided by persons suffering a rare genetic disease as well as by their family members. The sample donor has no contact with the biobank since the samples are provided through an intermediary such as a medical doctor. Therefore, the administrative procedure can be illustrated as follows. First, a blood sample together with some additional information (the name of the donor, the donor's date of birth, pathology) and the informed consent form arrives at the biobank. Both sample and additional information are provided by a so-called 'contributor', typically a medical doctor with his own research agenda. Then the following information is entered in a reception book: the reception date, the name of the donor, the donor's date of birth, and pathology together with the name of the 'contributor', state of health (e.g. *atteint* or *non-atteint du maladie*) and the nature of the sample (e.g. blood). Also, the sample receives a serial number (*code individu*) according to the entry into the reception book, and a family code (*code famille*). Hereafter, all data is fed into a database and a sample code (*code Généthon*) is generated by the computer system. In the moment this code is generated, name and surname of the donor disappear on screen: the sample is coded. The *code Généthon* is printed out on a sticker and stuck on the sample, which is then further processed. Thereafter, samples can be processed and stored in four forms: extracted DNA, total RNA, purified lymphocytes and established lympholastoid B lines.

One biobank employee explained that '[w]e are truly a service institution for research teams', although a *Généthon* interviewee argued that the biobank exists 'to help the sick persons'. In this regard, the *Généthon* bank's meaning is articulated as 'promot[ing] progress in genetic research in the interest of patients and their families, by making high quality cells and human products available to the scientific community' (Généthon, The DNA and Cell Bank. Online. Available www.genethon.fr, accessed 20 August 2005). Practically, this means that samples are provided to so-called 'users' who are typically (inter)national researchers or *Généthon* researchers. To the former, samples are provided if their research project seems promising for the cause of the patient organization and beforehand a contract is drawn up to regulate the details of the co-operation.[19] In other words, only storage and research that suits the research policy of the AFM will be supported. In practical terms, this means that each research team requesting samples from the *Généthon* bank has to issue a written request to the organization's *Arbitration Committee*, which consists of delegates of the AFM, the *Généthon* and the *Association Alliance des Maladies Rares*, and is assisted by experts where considered necessary. Furthermore, research teams are charged the material and shipping costs, and have to acknowledge in publications derived from the research that the results were based on samples from the *Généthon* biobank. These and other conditions of use are outlined in the biobank's charter (Généthon 2006) and represent the written manifestation of governance of biobanks (mode 1, see Gottweis and Petersen, Chapter 1). Since 2001, all users of the biobank have to act in compliance with the charter, which is periodically revised.

Conclusion

In this chapter, I have argued that biobanking is a practice that is open to articulation and rearticulation. Its meaning is context dependent and contingent and has to be negotiated in the broader context in which it is embedded. I have done so in the particular case of the *Généthon DNA and Cell Bank,* which is governed by the French patient organization AFM. I have shown that the AFM is an important stakeholder in both research endeavours and biobanking activities related to biomedical research in France. It articulates biobanking as a necessary means to advance biomedical research and operates against the culture of French etatism. Since the late 1990s, the AFM has initiated several biobanks and has shown itself to be a very active actor in the French biobanking scene. It has encouraged the CCNE's opinion on biobanks, fostered collections at hospital sites (e.g. *Banque Cassini*) and supported the creation of networks of biobanks[20] (e.g. *EuroBioBank*).

Among the many biobanks in relation to the AFM, I have presented the *Généthon DNA and Cell Bank* as the most intriguing case, insofar as the AFM retains control over it at all times in compliance with the 'partnership model'. Generally speaking, the degree of involvement of patient organizations in biobanking activities is varied, and ranges from the organization of money raising events for biobank projects to the participation in the governance of a particular biobank (mode 1), helping, for instance, to develop an informed consent form. Moreover, the patient organization governs through biobanks (mode 2). Clearly, the *Généthon DNA and Cell Bank* is quite unique in how it is embedded in biomedical research practices. Its corporate identity is determined by the goals and promises of the stakeholder institution, which is financially independent and fosters all kinds of genetic research by various means. Using the concept of 'biological citizenship', I argued that disease is a constituting element of an individual's identity as a patient. This individual might join or be assigned to a group, such as a patient organization, around a particular biological component. Donating blood to a biobank, for instance, can thereby be understood as a symbolic act, as an articulation of the refusal to accept the biological component of one's identity as a 'fate'. In an entrepreneurial manner, the act of donation can be depicted as an investment not only in a (possibly) 'better' future for one self, but also for all those who share the refusal to accept disease as a final verdict. Ultimately, I have argued that the *Généthon DNA and Cell Bank* exemplifies the operation of the 'political economy of hope' and thereby represents an institutional articulation of the concept of 'biological citizenship'.

Acknowledgements

To all my interviewees, especially to the staff members of the *Généthon DNA and Cell Bank,* I express my gratitude for their generosity and patience with a non-native French speaker. Also, I would like to thank my various reviewers, especially Hernàn Cuevas Valenzuela, for their comments on earlier drafts of this chapter, Alan Petersen for his editorial guidance, and Jean-Paul Gaudillière and Herbert Gottweis for their continuous support of my research and PhD project. The research was funded by the GEN-AU programme of the Austrian Federal Ministry for Education, Science, and Culture. Academically, it benefited from a three months stay at the French research institute CERMES in 2004. Conclusions drawn are my own, as are any errors.

Notes

1 I argue according to Laclau and Mouffe ([1985] 2001) that all objects, actions and practices are meaningful and are constituted as objects of discourse. Practices, they argue, have to be considered as discursive at all times. Thus, a distinction between discursive and non-discursive practices and discourse is not fruitful for our analysis as practices are always discursive.

2 For further distinction between governance of biobanks (mode 1) and governance through biobanks (mode 2), see Gottweis and Petersen, Chapter 1.
3 This chapter is primarily based on observations at French biobanks and a series of semi-structured interviews with key actors in the French biobanking field conducted intermittently between September 2004 and June 2005. The interviews were conducted in French and later translated by me into English. All extracts taken from interviews are anonymized.
4 Emily Martin, for instance, describes the body by way of an analogy with the nation state. In this sense, the body is similar to the 'nation state at war over its external borders, containing internal surveillance systems to monitor foreign intruders'. (Martin 1990: 410). In case of rare disease patients, their war is fought over internal borders because the 'enemy' is inscribed on the genetic level and might be called an unwanted life threatening and internal 'other'.
5 Rabeharisoa (2003) presents the 'partnership model' in contrast to two other models: the 'auxiliary model', and the 'emancipatory model'. In case of the latter, a patient organization tries to put its specific matter on the political research agenda (e.g. HIV/AIDS activists in the 1980/90s) and rearticulates the traditional roles between patients and scientists. In case of the former, decision-making power is delegated to a scientific council and the patient organization contains the role of a direct payer for the research conducted. The traditional roles are maintained.
6 The AFM and its activities, however, are not uncontested.
7 With the concept of 'etatism' I refer to the French political culture in which the state is expected to solve citizens' social and political issues (Kempf 1999; Christadler 1999).
8 In relation to biobanking, no single law is applicable to biobanks. Rather, there exists a whole network of legislation and regulations mostly addressed in the *Code Civil* (esp. Article 16), and the *Code de la Santé Publique*. Part of the latter is the *Law Bioéthique*, which was revised in summer 2004, and contains articles relevant for biobanking (e.g. informed consent, mandatory declaration of collections). Additionally, legislation for the protection of computerized files (responsibility: the *Data Protection Supervisory Authority*, CNIL) and the requirements for the national support programmes 'cohortes et collections' (responsibility: *Ministry of Research* and Inserm) and 'tumorothèque' (responsibility: *Ministry of Health*) apply to and shape biobanking activities.
9 It was furthermore part of a dispute, a scandal, concerning 'French DNA' portrayed by Paul Rabinow 1999.
10 The participation in the support programmes was voluntarily. By 2006, about 100 biobanks showed interest and forty-seven biobanks were selected for funding. For further information about France, BRCs, the restructuration and reevaluation of exisiting collections see www.crb-france.org (accessed 6 October 2006) and *La Lettre de la Délégation à la Recherche Clinique d'Ile de France* 2006, 2. Further research is needed to truly assess to what degree the BRC initiative is of importance for France.
11 In 2000, the *Laboratory of Medical Ethics and Public Health at the Faculté de Médecine Necker* (*Université Paris V*) published a study that evaluated biobanking activities in several departments of the *Necker University Hospital* and the *Reims Hospito University*. Their survey, which draws upon questionnaires that were sent out to sixty departments of which thirty-three were returned, concludes that '20 [departments] possess a DNA collection and returned a completed questionnaire; 11 do not use a collection and did not complete the questionnaire; and 2 departments replied a letter explaining that they did not fully agree with the definition of a DNA bank given in the introduction [. . .]' (Moutel *et al.* 2000).

12 In the French context, the terms biolibrary (*biothèque*) or biorepository are synonymosly used for biobank. So far, there is no legally binding definition of the term. For further information on this topic see Sandrine Caze de Montgolfier (2002) and CCNE (2003).
13 The AFM is contested by many. The *Syndicat National des Travailleurs de la Recherche Scientifique* (SNTRS), for instance, argues that the patient organization's research policy leads to a privatisation of research and a disengagement of the French state in public research (SNTRS 2003). The SNTRS's criticism exemplifies the viewpoint that the state rather than private initiatives should foster and retain control over basic research.
14 In 1989, the AFM started to finance gene therapy. In 1998, it launched a scientific research programme called the 'Great Attempt' that led to the development of tools, networks and research centres for gene-based therapies (AFM, From the 'Great Attempt' to the 'New Frontier'. Online. Available www.afm-france.org, accessed 27 September 2006).
15 *Généthon* develops therapeutic products for the treatment of rare diseases and is also involved in gene therapy.
16 *Genopole* was created by an initiative of the French government and the AFM in 1998. Today, it comprises twenty research laboratories and about forty biotech companies.
17 Here, I do talk about the corporate identity of the biobank. I do not talk about how individual employees identify with the actions of the AFM or *Généthon.*
18 The *Bank On A Cure* is located at the *University of Minnesota Cancer Centre* in Minneapolis, USA, whereas the headquarters of the IMF is in California.
19 A so-called 'contributor' is a person who has initiated the collection of samples that are registered and stored at the *Généthon DNA and Cell Bank.* This person is usually a medical doctor or researcher and has the right to oppose all access to the samples by a third party (Généthon 2006).
20 For further reading on networks of biobanks see Mayrhofer and Prainsack 2008.

References

AFM (n.d.) DNA collection, available online at www.afm-france.org (accessed 20 August 2005).

AFM (n.d.) 'From the "Great Attempt" to the "New Frontier", available online at www.afm-france.org (accessed 27 September 2006).

AFM (n.d.) 'Our mission', available online at www.afm-france.org (accessed 27 September 2006).

AFM (2004) 'Rare diseases. From recognition to treatment: role of the AFM', available online at www.afm-france.org/afm-english_version/e_upload/doc/rare_diseases.doc (accessed 13 August 2005).

Barataud, B. (1999) *L'Effet Téléthon.* Paris: Les Essentiels Milan.

Brown, N. and Webster, A. (2004) *New Medical Technologies and Society. Reordering Life.* Malden, MA: Polity.

Cambon-Thomsen, A. (2003) 'Assessing the impact of biobanks', *Nature Genetics*, 34: 25–6.

Cambon-Thomsen, A. (2004) 'The social and ethical issues of post-genomic human biobanks', *Nature Reviews Genetics*, 5: 866–73.

CCNE (2003) Opinion No. 077. 'Ethical issues raised by collections of biological material and associated information data: "biobanks", "biolibraries", available online at www.ccne-ethique.fr (accessed 20 November 2004).

Christadler, M. (ed.) (1999) *Länderbericht Frankreich: Geschichte, Politik, Wirtschaft, Gesellschaft*. Opladen: Leske + Budrich.

Clarke, A. E., Mamo, L., Fishman, J. R., Shim, J. K. and Fosket, J. R. (2003) 'Biomedicalization: technoscientific transformations of health, illness, and US biomedicine', *American Sociological Review*, 68: 161–94.

Cohen, D. (1993) *Les gènes de l'espoir: à la découverte du genome humain*. Paris: Robert Laffont.

Crawford, R. (1985) 'A cultural account of "health": control, release, and the social body', in J. B. McKinlay (ed.), *Issues in the Political Economy of Health*, New York: Methuen-Tavistock.

CRB (n.d.) 'CRB', available online at www.crb-france.org (accessed 6 October 2006).

de Montgolfier, S. C. (2002) 'Collecte, stockage et utilisation des produits du corps humain dans le cadre des recherches en génétique: état des lieux historique, éthique, et juridique, analyse des pratiques au sein des biothèques', Paris: Université René Descartes Paris V. Available online at www.ethique.inserm.fr/inserm/ethique.nsf/0f4d0071608efcebc125709d00532b6f/933a7f5b352f398ac12570a500515255/$FILE/_2ahingt3541kmst42ctp62r0_.pdf (accessed 20 November 2004).

Généthon (n.d.) 'The DNA and cell bank', available online at www.genethon.fr (accessed 20 August 2005).

Généthon (2006) 'DNA and Cell Bank Charter', available online at www.genethon.fr/fileadmin/user_upload/Redacteur/Pdf/05_Banque_ADN/Bank_Charter.pdf (accessed 14 August 2007).

Godard, B., Schmidtke, J., Cassiman, J.-J. and Aymé, S. (2003) 'Data storage and DNA banking for biomedical research: informed consent, confidentiality, quality issues, ownership, return of benefits. A professional perspective', *European Journal of Human Genetics*, 11: 88–122.

Hirtzlin, I., Préaubert, N. and Charru, A. (1999) 'Analyse de l'activité et du coût des collections de matériel biologique', *Journal d'Economie Médicale*, 17: 3–16.

Hirtzlin, I., Dubreuil, C., Préaubert, N., Duchier, J., Jansen, B. *et al.* (2003) 'An empirical survey on biobanking of human genetic material and data in six EU countries', *European Journal of Human Genetics*, 11: 475–88.

Kaufmann, A. (2004) 'Mapping the human genome at Genethon Laboratory: the French Muscular Dystrophy Association and the politics of the gene', in H.-J. Rheinberger and J.-P. Gaudillière (eds), *From Molecular Genetics to Genomics: The Mapping Cultures of Twentieth Century Genetics*. London: Routledge.

Kempf, U. (1999) *Das Politische System Frankreichs: Eine Einführung*. Opladen: Westdeutscher Verlag.

Kent, A. (2002) 'Patients + research = result! The role of patients and their interest groups in biomedical research', *EMBO Reports*, 3: 707–8.

Knoppers, B. M. (ed.) (2003) *Populations and Genetics. Legal and Socio-Ethical Perspectives*. Leiden, Boston, MA: Martinus Nijhoff Publishers.

Laclau, E. and Mouffe, C. ([1985] 2001) *Hegemony and Socialist Strategy. Towards a Radical Democratic Politics*. London: Verso.

Latour, B. (1988) *The Pasteurization of France*. Cambridge, MA: Harvard University Press.

Martin, E. (1990) 'Towards an anthropology of immunology: the body as nation state', *Medical Anthropology Quarterly*, 4: 410–26.

Mayrhofer, M. and Prainsack, B. (2008) 'Being a member of the club. Transnational networks of biobanks', *International Journal of Risk Assessment and Management*, forthcoming.

Medizinische Universität Graz (2006) 'Entwicklungsplan der Medizinischen Universität Graz, 2006020', available online at www.meduni-graz.at/leitbild/pdf/entwicklungsplan.pdf#search=%22biobank%20und%20pathologie%20graz%20sammlung%22 (accessed 29 September 2006).

Moutel, G., de Montgolfier, S. and Hervé, C. (2000) 'Biobanks and DNA banking: an analysis of the practices in 20 hospital departments', available online at www.inserm.fr/ethique (accessed 20 November 2004).

Nationaler Ethikrat (2004) 'Biobanken für die Forschung. Stellungnahme', available online at www.ethikrat.org/themen/pdf/Stellungnahme_Biobanken.pdf (accessed 1 October 2006).

Parsons, T. (1951) *The Social System*. New York: Free Press.

Petersen, A. (2005) 'Securing our genetic health: engendering trust in UK Biobank', *Sociology of Health and Illness*, 27: 271–92.

Pontille, D., Milanovic, F., Rial-Sebbag, E. and Cambon-Thomsen, A. (2006) 'Le vivant à l'epreuve des collections: biobanques et bases de données en recherche biomédicale', Centre d'Etude et de Recherche Travail, Organisation, Pouvoir CERTOP-UMR CNRS 5044 & Inserm U 558 Epidémiologie et Analyses en Santé Publique. Rapport Final.

Rabeharisoa, V. (2003) 'The struggle against neuromuscular diseases in France and the emergence of the "partnership model" of patient organisation', *Social Science & Medicine*, 57: 2127–36.

Rabeharisoa, V. and Callon, M. (1999) *Le pouvoir des malades. L'association française contre les myopathies et la recherche*. Paris: Les Presses de l'École des Mines.

Rabinow, P. (1999) *French DNA. Trouble in Purgatory*. Chicago, IL, London: University of Chicago Press.

Rose, N. and Novas, C. (2005) 'Biological citizenship', A. in Ong and S. Collier (eds), *Global Assemblages. Technology, Politics, and Ethics as Anthropological Problems*. Malden, MA: Blackwell.

Strauss, A., Corbin, J., Fagerhaugh, S., Glaser, B., Maines, D. *et al.* (eds) (1984) *Chronic Illness and the Quality of Life*. St Louis, MO, Toronto: C. V. Mosby.

Sundström, C. (2001) 'The biobank resource in anatomical pathology', in M. G. Hansson (ed.), *The Use of Human Biobanks. Ethical, Social, Economical and Legal Aspects*. Uppsala: Uppsala University.

Waldby, C. (2002) 'Stem cells, tissue cultures and the production of biovalue', *Health: An Interdisciplinary Journal for the Social Study of Health, Illness and Medicine*, 6: 305–23.

Waldby, C. (2006) 'Umbilical cord blood: from social gift to venture capital', *BioSocieties*, 1: 55–70.

Waldby, C. and Mitchell, R. (2006) *Tissue Economies. Blood, Organs, and Cell Lines in Late Capitalism*. Durham, NC, and London: Duke University Press.

Williams, G. and Schroeder, D. (2004) 'Human genetic banking: altruism, benefit and consent', *New Genetics and Society*, 23: 91–103.

6 'This is not a national biobank . . .'

The politics of local biobanks in Germany[1]

Ingrid Schneider

Biobanking is a globalized phenomenon that was triggered by scientific and technological advances in the wake of the Human Genome Sequencing Project. Population-based biobanks in the life sciences articulate the shift from rare, monogenetic diseases towards common diseases of high prevalence (Holtzman and Marteau 2000; Halliday *et al.* 2004). This shift is driven by disciplinary and professional research strategies which aim to expand the explanatory power of genetics, and thus entail the 'geneticization of medicine' (Lippman 1991).

The nation state seems to play a great role in enabling population-based genetic databases. 'National' biobank projects in Iceland, Estonia and the UK have caught international attention. In contrast, biobanks in Germany primarily appear as local practices. My analysis is centred on the question: Why, as yet, is there no large-scale, single, 'national' biobank project in Germany? This inquiry ad negativum points to the structural conditions for the establishment of population genetics databases, and to the interaction between science and society; it does not presume that biobanks must necessarily be modelled in a centralized, 'national' fashion. As I will explore, there is a special politics behind this local framing that can only be adequately investigated if historical and social trajectories are taken into account. Therefore, the constitution processes of 'locality' and the special functions of 'localization' in the context of national and global biobanking activities form important points of reference for my analysis.

As I argue, there are several determinants for establishing biobanks at the national level. Among them are institutional structures of the state (centralism or federalism), of research funding, and of the health system (private/public), which means that path-dependencies do matter. Furthermore, some historical, cultural, and symbolic dimensions of biobanking are explored that are related to notions and correlations between the individual bodies of citizens and the 'nation's body'. These trajectories impact on 'naming' and 'labelling' processes, which are not merely rhetoric exercises. The anticipation of the public's reaction provides one of the clues for explaining the 'localized' governance model as pursued in Germany.

After having presented some politico-structural and socio-cultural prerequisites for the establishment of population-based genetic databases, the second part of my analysis explores the governance *of* biobanks (see Gottweis and Petersen, Chapter 1). In this respect, governance is being performed, first, by pre-existing monitoring bodies for clinical research and data protection; as so far no specialized regulatory authority is in sight, recommendations by the National Ethics Council have come to play a major role, and will be subjected to a critical analysis. Second, governance is executed via the (re)interpretation, transformation, and enforcement of medico-legal rules, and third, by self-regulatory activities of biobank actors and networks which go beyond the local and national level.

The third part of this chapter deals with governance *through* biobanks. It investigates different visions on the impact on medicine and the health system; these contrasting forecasts and the way they are socially communicated will be shaping the interface between science, technology and society, and thus also the possible futures of biobanks themselves.

Biobanks: research funding, scientific demands, and structures of the health system

Germany to date has not started a centralized, nationwide biobank project. However, many biobanks have been set up at local hospitals, public research centres or in non-profit institutions. According to a recent survey, there are eight larger biobank projects and around thirty disease-related biobanks in Germany (TMF 2005). Some biobanks started as public–private partnerships.[2] Biobanks for pharmacogenomics are being established by several pharmaceutical companies. Most of the resources for the establishment of biobanks have been provided by the Federal Ministry of Education and Research (BMBF) for health-related research programmes, particularly to the 'Competence Networks in Medicine' (KN), established and funded since 1999, and the National Genome Research Network (NGNF), which was founded in 2001. Federal research funding supports the establishment of networks both of basic and clinical research centres to create synergies. Grants for biobanks as platforms for these networks are predicated on the assumption that research on the genetics of common diseases could contribute to the health, longevity and productivity of the population, trigger scientific and technological innovation, and enhance the economic competitiveness of the country (BMBF 2002, 2003; Schreiber 2003; Wagenmann 2005). Genetic epidemiology for common, complex diseases requires large population-based sample collections. As a principle it is stated that the smaller the effects of gene-environment-interaction, the bigger must be the size of the population studied (Roos 2001).

Scientific researchers in Germany are very sceptical about any large-scale approach in biobanking. Most of them think that only smaller, well-defined, disease-group specific projects, in which sophisticated phenotyping can be guaranteed, will yield valid outcomes. They point to the importance of extremely well-defined phenotyping, which requires reliable medical diagnosis, and specific questionnaires about the issues and data of relevance. Both are crucial in terms of the quality and validity of the scientific research. What is emphasized is that even a small amount of 'bad phenotyping' may 'spoil' the data in terms of valid correlations between genotype and phenotype structures, and thus may provide statistical artefacts with defective or even useless results: 'The databases will only be as good as the individual clinical and exposure information that they contain' (Wichmann *et al.* 2005: 586). Therefore the 'catch-all approach' pursued for instance in the UK or in Iceland is viewed with scientific reluctance. But there are also other reasons for choosing another approach, which are related to the structures of the health and insurance systems.

To capture a totality of a disease-specific or territorially defined population for research, access to patients' data is needed. Nations with a national health system have a comparative advantage in this respect, and therefore, large biobank projects emerged in countries like the UK, Canada, or the Scandinavian countries. In Germany, there is no such unified system, although almost complete health coverage is being achieved. The German health system is a mixed private–public, and multi-level system. 90 per cent of the inhabitants are covered by more than 250 mandatory statutory health insurances, the remaining 10 per cent by private insurances. The federalist structure of the state is decisive for most of the health-related issues. Most hospitals are owned and administrated by the Länder or municipals, although the current trend favours privatization of hospitals. Free choice of medical doctors is the norm, 'doctor hopping' a widespread practice. Specialist physicians operate both at the private, ambulant level and at the public, hospital level. Financing of private physicians is organized in a complex system mediated by professional associations at the Länder level. As a result, there is neither a centralized health data registry, nor are there coherent, accessible patients' databases which forms an obstacle for population-based biobanking. To sum up, institutional structures of the health system do matter for the establishment of biobanks. In the following paragraphs, I will concentrate on the two major, federally funded biobank projects.

The PopGen Biobank in Schleswig Holstein

The largest genetic biobank project in Germany is PopGen, based at the Kiel University Hospital. PopGen is situated in the northern part of the state Schleswig Holstein, an area comprising 1.1 million inhabitants, 41 hospitals and 1,700 private practitioners. One of the reasons for choosing this area was its geographic confinement at the border to Denmark which causes most people to use the local medical system. According to PopGen this situation allows for identifying all locally prevalent cases for the disease in question and thus achieving unbiased sampling. PopGen to date has targeted recruitment of all the patients in the area which are suffering from sixteen predefined disorders. Collection is performed on behalf of the national neuropsychiatric, cardiovascular, and environmental disease competence networks. Patients are enrolled at hospitals, via private practitioners, and through data provided by health insurers. Additionally, a representative control group of 7,500 randomly selected – presumably – healthy volunteers is being recruited. Inclusion criteria are sex and age (born between 1923 and 1988), and no other data are recorded (Krawczak *et al.* 2006).[3] For its phenotype databases, PopGen is gathering clinical data on the individual medical history, other diseases, medication, lifestyle, and family patterns. As yet, DNA samples from 45,000 participants, including the control cohort, have been stored. Further collection of at least a similar number of samples is expected, the number of diseases will also be expanded.[4] At present, PopGen's objective is not to identify new genes correlating to certain diseases. Its goal is to validate candidate genes detected in studies of patients with a strong family history. These genetic variants are tested for in a 'representative', 'normal' population, and for penetrance factors and relative genotypic disease risks in 'average' disease populations. Therefore, PopGen characterizes itself as a 'verification and validation machinery' for candidate genes.[5]

The KORA-gen Biobank in Augsburg

The KORA ('Cooperative Health Research in the Augsburg Region') study represents a population-based cohort of 18,000 adults in a region of 600,000 inhabitants. KORA is the follow-up project of MONICA ('Monitoring of Cardiovascular Diseases'), initiated by the WHO and conducted since 1984 in twenty-five countries in Europe, Australia and North America with standardized protocols. This was the first multinational attempt to correlate the incidence of diseases with known risk factors, such as personal lifestyle, quality of healthcare, and economic conditions (www.gsf.de/kora-gen). KORA is based on four surveys of 4,000–5,000 participants each, aged 25–74 years, which were followed up. For genetic epidemiological research based on the KORA cohort, KORA-gen has been established as a resource, which provides controls for genetic studies, DNA, plasma, serum samples, and immortalized cell lines. Phenotypical information is available on cardiovascular diseases, type 2 diabetes, stroke, cancer, asthma/allergies, and general health status (Wichmann *et al.* 2005: 587). Data and services are supplied for external use by the national genetic and disease networks, and other research groups, according to certain rules, encompassing scientific co-authorship, and payment of fees (Holle *et al.* 2005).

Future developments: a 'Biobank Virtual' or a 'Biobank Central'?

For the question of why there is no 'Biobank Germany', at this point, a first response can be given. Implicitly, there are two competing concepts for a 'national' biobank project. KORA favours a 'virtual' biobank, which integrates several existing epidemiological cohorts into a common pool (Wichmann *et al.* 2005: 588). Six population-based cohorts would sum up to more than 127,000 research subjects. This model derives from risk epidemiology and strives to accommodate genetics. In contrast, PopGen may tacitly aim to become *the* 'central' biobank for German genetic research (Schreiber 2003: 19). The many disease-specific samples accrued by PopGen can over time reach a critical size which would allow for the statistical power to embark on genetic association studies or data mining strategies. The tension between these two models is latent and has so far not resulted in conflicts over the allocation of federal research funds. In the future, both strategies may even complement each other. However, should more big countries pursue a large-scale, centralized population-based biobank[6], this might induce politicians to adopt the same policy. For the strategy mapped out by German researchers, the difference between centralized or federalist political systems is also decisive. Had researchers from the beginning put a 'German Biobank' on the agenda, tricky negotiations between the sixteen Länder would have followed over who would get the federal resources, and furthermore, how to strike a fair East–West balance. Should the upcoming international *Zeitgeist* call for a central genetic population project, then PopGen's self-representation may shift from understatement to exuberance, in suggesting that it already *is* the large-scale biobank project which supposedly forms an indispensable asset for any innovative nation in the age of knowledge-based economies.

Why 'local' biobanks? The role of the public

Explanations for why biobanking in Germany is primarily framed as a *local* endeavour must also consider the role of the public. To be successful, funding institutions' and researchers' portrayal of their projects has to anticipate public reaction. Germany has a highly politicized conflict culture regarding privacy and genetic research. There was massive resistance towards a population macro census in 1983 in West Germany, which subsequently failed and consequently no macro census has been performed ever since. In this case, opposition was also brought to the Federal Constitutional Court (*Bundesverfassungsgericht*), resulting in a strong, constitutionally backed 'personality right for informational self-determination'.[7] This right involves control over what personal information is collected or generated, and control of disclosures to third parties. More recently, this 'personality right' has been spelled out as 'right to bio-informational and bio-material self-determination' (Halàsz 2004). The constitutional court decision has set high standards for privacy, which were institutionally backed by the establishment of data protection laws and officers (agencies) both at the Federal and at the Länder level.

Opposition towards genetic research was articulated on many occasions and there have been broad public debates on biotechnology and biomedicine. As a result, biotechnological safety was regulated in the Genetic Technology

Act in 1993. The Embryo Protection Act in 1990 provided restrictive statutory legislation in relation to IVF treatment. Since 2002, the importation of human embryonic stem cells has been governed by the Stem Cell Act, which is perceived by many scientists as restricting their 'research freedom'. Broad public opposition to the 'bioethics convention' of the Council of Europe, which was seen as undermining strict national standards in biomedicine, has led to non-ratification by the government.

Therefore, any framing of biobanks as a 'national' endeavour would call for federal statutory regulation. The term 'national' would be translated as of 'common, national interest'. By the same token, any suggestion of a national 'German Biobank' might alert national media attention, general sensitivities, and possibly provoke public outcry and the resistance of NGOs. Any 'national' biobank might furthermore arouse fears of a 'national genetic registry' and thus of state surveillance and control; in this respect, there are high sensitivities in the Eastern part of Germany – the former German Democratic Republic. 'National' biobanks in Iceland, Estonia, and the UK have received very critical media attention in Germany. In contrast, 'local' biobanks are regarded as being of only 'local' relevance, especially if performed in predominantly rural areas (Schleswig-Holstein – PopGen) or in 'minor' cities (Augsburg – KORA-gen), which are less prone to popular protest. For actors who do not seek political furore, 'localization' of biobanks has meant *depolitization*. However, the 'local' framing is misleading insofar as 'localization' refers only to the recruitment area of biobanking projects, whereas the use of data and samples is taking place at the national – and possibly in the future – on the international level.

Biobanks and national identity

Narratives of a common ancestry and a society's identity which can be told with a certain credibility and pride have lent support for the establishment of national biobank projects. While Iceland's deCODE company has invested heavily into the Viking myth, one of Estonia's main narratives was to 'put the country on the world's map'. The UK Biobank alike has invoked national pride. In Germany, no such a narrative is available. Germany's history is fraught with a belated nation-building process and the violent abuse of (hyper)nationalism in the 'Third Reich'. Difficult nation-building processes are continuing until today after the re-unification of East and West Germany. The country has never attained a positive self-image of '*grande nation*', like France has. Germany is proud for being a *'Kulturnation'* and still refers to itself as the country of Goethe and Schiller. There is also a strong post-war tradition of being a '*Rechtsstaat*', a state based on the power of the law, which conceptualizes the public interest in abstract, legal, and unitary terms (Jasanoff 2005: 7). What is paramount as one of the major determinants of Germany's political identity is '*Verfassungspatriotismus*' (constitutional patriotism), a normative notion invoked by the philosopher Jürgen Habermas

(1986: 75). A 'Biobank Germany' would conjure up the dark shadows of National Socialism, of 'medical' atrocities, and of 'racial hygiene' as an eugenic ideology and political programme.

Supposedly, in any 'national', centralized biobank project, images of the *nation's* body and identity are linked to the collections of parts of *individual* bodies. As *pars pro toto* biobanks symbolize repositories of the national population and its bio-social heritage. In Germany, however, even the term 'collective body' would be rejected because of the historically tainted NS-ideology of a (unified) '*Volkskörper*' (body of the nation). Any mention of 'Germanic genes' or relationships between genetic heritage and territoriality may recall NS-'blood and soil' ideology that would immediately be repudiated. This is not to say that Iceland's biobank has ever claimed to find 'Icelandic genes'. But its narrative invoked genetic *homogeneity* (as represented by the Vikings), and therefore was presented as a viable way to 'find the genetic needle in the haystack', and genetic '*representativeness*', at least for the world's Caucasian population and its ailments. Both narratives would not work for Germany: genetic homogeneity would not only be refuted because of centuries of immigration, but moreover, would point to the elimination of Jewish people and to the pursuit of 'genetic purification' strategies, which led to social exclusion and ultimately to the Holocaust. This perception may even be mirrored from the outside: Would anybody regard the German population as another country's genetic 'representative'? Thus there is no positive, credible metanarrative, on which a German '*national*' biobank could draw. 'Germanic' genes remain an unthinkable relationship as they would immediately recall notions of 'aryan supremacy' and subsequent selection and extirpation strategies.

Legitimate narratives: local patriotism and transgenerational solidarity

While nationalism is not available as a narrative for genetic biobanking, 'local' biobanks can (and do) draw on local patriotism and pride. Local patriotism is not only allowed in Germany, but also highly appreciated. By organizing a 'Parliamentary Evening' at the state parliament, attended by more than eighty politicians, support from local politicians for the PopGen project was generated during its incubating phase. Loyalty and pride are associated with PopGen as a major scientific project, which is seen as a potential job creator and instrument for the funnelling of federal research funds into the Länder. The national media's coverage of biobanks within Germany is relatively low and mostly descriptive. Local newspapers are in favour of PopGen. PopGen itself has employed a sociologist and PR specialist who devotes an important part of her time to contacts with media, politicians, scientists, potential donors, and members of the region's healthcare system. Critical voices of physicians, which were strongly articulated in Iceland, are virtually absent in Germany.

Enrolment to the biobank's studies is motivated by invoking transgenerational solidarity: PopGen's main slogan is 'health for generations'. Leaflets are alluding to reciprocity and charity, but also to a duty for health, by 'citing' research participants suffering from diseases: 'I want to help where I can, because a healthy life is even more beautiful, because I love life.' Parental obligations towards the offspring are mobilized by the quote 'I want to help to keep my children healthy'. Another legitimate narrative is care for the community and contribution to the 'common good', both primarily played out as cosmopolitan solidarity. To sum up, a localized framing of biobanking allows for suppressing historical sensitivities and taming of political attention which would be associated with any 'nationalized' approach, while it mobilizes local patriotism and pride. After having identified some characteristics associated with the establishment and public representation of population-based genetic biobanks, in the next section I will consider how these are governed.

Regulatory frameworks and monitoring bodies

Biobanks do not operate in a zone free from legal regulations. Quite the contrary. There are many layers of national laws – medical, administrative, institutional, property in rem, intellectual property, professional, civil, criminal, labour, confidentiality, constitutional law – which are implicated in the practices of biobanking (cf. Simon *et al.* 2006). EU regulations (e.g. on data protection, databanks, drug legislation), and international norms governing medical research (Nuremberg Code, Declaration of Helsinki) must also be adhered to. However, these overlapping frameworks are contradictory, and there is as yet no authoritative guidance on how to interpret and balance the different norms, rights, and obligations, as they are not specified for biobanks (Cambon-Thomsen 2004). At present, the two most important German regulatory bodies for the governance of biobanks are the fifty-five ethics commissions (institutional review boards) which have to be consulted prior to research on human subjects to give a favourable vote, and data protection officers who monitor compliance with confidentiality and data protection laws. Both regulatory bodies primarily operate at the level of the Länder or the university hospital.

Governance by expert commission: the National Ethics Council's opinion on Biobanks

In a situation fraught with legal uncertainties, an opinion paper by the National Ethics Council (NEC 2004), an expert advisory commission for the German government, has become a major point of reference for the emerging governance of biobanks. NEC's opinion paper was enthusiastically received in the international biobanking's epistemic community: It was hailed as 'refreshingly innovative and progressive', exhibiting an 'enlightened, pragmatic and practical approach' (Knoppers 2004). German researchers welcomed it as helpful and 'research-friendly' (Wichmann *et al.* 2005: 586).

In its regulatory recommendations, NEC states that the 'central element of all proposals must be the donor's right of self-determination' (NEC 2004: 11) which is spelled out as voluntary and informed consent. However, the informed consent prerequisite in the NEC's opinion paper is paradoxically constructed: on the one hand, it is supercharged as the incarnation of research's legitimacy. On the other hand, *informed* consent is hollowed out, because of suggesting *generalized* consent to *all* possible, future, indeterminate, timely unlimited uses for medical research of individual samples and data, without requiring any re-contacting and re-information mechanisms. Thus it is culminating in *blanket* consent, devoid of any information on specified purposes of storage, research and use (NEC 2004: 13; 60). The '*clou'* in the recommendations consists in the suggestion to exert individual autonomy by relinquishing individual control.

The option of withdrawal can be considered as the only 'emergency brake' left to the research subject. It entails an individual veto on further participation and is regarded as a right which cannot be relinquished[8] (NEC 2004: 61). However, this right is practically diminished if there are no corresponding obligations for information, transparency, and accountability on the current use of samples and data. In practical terms, withdrawal is only possible in the absence of anonymization. For re-identifiable (coded) samples and data, withdrawal relates only to future uses, not to research data already aggregated (NEC 2004: 61).

Concerning the secondary research use of samples and data derived in the clinical context, NEC states that informed consent can be waived under certain conditions (NEC 2004: 11–12). Transfer of data, samples, or the entire biobank to third parties is deemed to be possible if some safeguards are met (e.g. codification or ethics commission's approval) (NEC 2004: 13).

In a third reading, informed consent can even possibly be regarded as ambiguous, and overly fraught. Information on potentially disadvantageous effects of research participation – the NEC opinion here mentions state sequestration of samples and data – can also be interpreted as shifts in responsibility from the researcher towards the individual research subject. It may indicate a transfer of liability: 'You knew what you did when consenting, then you should not complain about, or sue for adverse consequences.' Concerning financial compensation and commercial use, including patents filed, the informed consent is usually constructed as a disclaimer, i.e. a waiver or transferral of potential individual demands on ownership, co-inventorship, and financial benefit-sharing (cf. Simon 2006; TMF 2006; Goebel 2006). Thus it implicitly entails a *contractual* relationship in which the individual research subject is conveying rights on use, control, and profit-sharing from downstream research results. The terms of the contract are unilaterally defined by the research institution (cf. Winickoff 2003: 209). Undesirable socio-political implications – such as discrimination through disclosure requests by insurers and employers, and risks of social stigmatization of family members, patients' or ethnic groups – are, according to the NEC, not to be part of informed consent forms (NEC 2004: 68–70).

To sum up, the NEC's opinion regarding broad, generalized consent and waiver stipulations can be considered 'breakthrough' for the governance of biobanks. Should these recommendations be translated into statutory law, this would not only enable large-scale, prospective, multi-purpose, and long-term genetic biobanking practices. It would also abolish the hitherto legal or ethical prohibition of such 'blanket' consent, and the long tradition of purpose-bound informed consent requirements in medical research. Moreover, it would be a conceptual break with the normative tradition of individual rights taking precedence over the public good in liberal law systems. In this respect, the NEC statement seems to be customized to the needs and interests of the emerging genetic epidemiological research community. Although donors' self-determination is coined as central for NEC's regulatory proposals, empowerment of donors as a collective entity, or as stakeholders, is not provided for. As regulatory bodies, the NEC predicates all of its proposals on favourable votes of ethics commissions, where donors are not represented, and the oversight of data protection officers. Those monitoring bodies are deemed as providing substantive as well as procedural protection, as being both necessary and sufficient, unless the samples and data are fully anonymized. Technical means of data protection (coding or double-coding, encryption) are given precedence over regulatory and/or legislative governance. Participatory or discursive regulatory measures, such as mandatory consultation pre-, inter- and post-research with patients' advocacy organizations and the public are not even considered (cf. Winickoff, Neumann 2005; Schneider 2002a). Special transparency and accountability mechanisms (Williams 2005) for research performed on collections which can be electronically accessed, exchanged and pooled across networks are not proposed as public policy. In fact, donors are primarily treated as sources, not as citizens or participants.

Thorny policy questions of ownership and access to biobanks, such as non-exclusive and exclusive use of material as well as data, priority rules, and possible tensions between public, non-profit, and commercial institutional forms (Schneider 2002b) are only slightly addressed (NEC 2004: 19). Similarly, complex issues of Intellectually Property Rights, either in relation to inventions and industrial applications made as a result of research, or in relation to the design and content of databases, are nearly completely elided, except for being part of the informed consent requirements. Nonetheless, biobanks need to clarify their policies regarding management of Material Transfer Agreements, Intellectual Property Rights (copyright, patents, databases), and commercial aspects.

Mitigating public policy measures proposed by the National Ethics Council

However, the NEC opinion entails some innovative provisions that could be considered as complementary regulatory requirements for the protection of donors and the public good, which have not received the same international attention. I will shortly highlight some of these proposals:

- Appointment of a national *curator or trustee* for the supervision of biobank operations, and for monitoring of compliance with ethical standards and legal requirements.[9] This body 'should therefore be responsible for ensuring, for example, that donors' expectations (. . .) are complied with; that the relevant conditions of access to the biobank are observed; that the limitations on the transfer of materials or data set by the research vocation of the biobank (. . .) are not exceeded; and, finally, that if the biobank is closed down, its stored bodily substances and information are not misused' (NEC 2004: 64).
- The establishment of *'research confidentiality'* as a legal instrument to keep data confidential vis-à-vis state authorities.[10] This regulatory proposal addresses the potential of sequestration of samples and data stored in biobanks by police or state prosecutors as used for evidence seeking, to safeguard important public interests, to avoid dangers to national security or threats to the public welfare (cf. Simon *et al.* 2006: 153). The NEC called for a statutory requirement of confidentiality of research, 'prohibiting any use whatsoever for non-research purposes', arguing that donors' trust and willingness, and 'hence ultimately the acceptance of biobanks' would depend crucially on the certainty of non-use for any other than scientific research purposes (NEC 2004: 67). However, this compulsory confidentiality of research to be enshrined in law must also 'call for the striking of an appropriate balance, especially as regards to the possibility of access to data for the investigation of serious criminal acts' (NEC 2004: 68).
- *Collective benefit sharing* schemes are discussed as a response to economic gain accruing from the subsequent exploitation of research results. Although mandatory benefit-sharing mechanisms were not endorsed, voluntary contributions of profit-oriented biobank users to public-welfare funds are deemed as 'desirable' (NEC 2004: 20). Different models for such funds, either project-, disease-, or group-related, and to be established at the national or international level, are presented. What is also mentioned is the use of such funds for collective professional representation of donor's interests and rights, and health consumer advocates (NEC 2004: 77–8, cf. Schneider 2002a; TAB 2007).

These proposals for public policy instruments deserve more general discussion. Moreover, it could be argued that population-based biobanks can only be justified if such policy safeguards are introduced as *conditio sine qua non* for their establishment and maintenance. Furthermore, research collection of samples, and generation of large volumes of data give rise to particular risks of linkage, both to the individual and the citizenry as a whole; therefore they have to be accompanied by statutory law which prohibits genetic discrimination of individuals and groups by insurers and employers.

National statutory regulation and professional self-regulation

For the time being, legislative governance of biobanking, as has occurred in Iceland, Estonia, Sweden, Norway and Denmark, is not on the horizon in Germany. However, the German Parliament in 2005 commissioned its Office for Technology Assessment (TAB) to give advice on the need for specialized statutory initiatives. The TAB report recommends a 'biobank act' modelled according to a recent Swiss legislative proposal (www.samw.ch) as a policy option (TAB 2007). The scientific research community is currently not in favour of a 'lex biobank' which might provoke strong political discussions. However, broad permissive clauses for genetic research were introduced into a draft for an 'act regulating genetic diagnostics' elaborated by the Federal Ministry of Health. In its Chapter 7, termed 'genetic investigations for purposes of scientific research', recommendations of the NEC opinion on biobanks were taken up, sometimes even verbatim. This applies to the permissibility of generalized consent and waiver stipulations, rules regarding anonymization and 'pseudonymization', validation by ethics commissions, rights to information of the affected persons, storage and deletion of samples, the publication of results, and with regards to persons in need of special protection, such as minors and incapacitated persons (BMGS 2004, §§26–33). Should this draft act on genetic diagnostics be introduced and finally passed by the Parliament, this would provide 'generous' legislative conditions for research. The 'backdoor' approach chosen would mean legal certainty for the researchers without arousing public political noise. Regulation would thus be more enablement than restriction, be useful to limit researchers' liability, and prevent possible litigation initiated by sample donors. Whether such statutory rules would pass the Constitutional Court test, however, remains uncertain.

But even if this legislation were passed, many other important aspects of biobanking would not be regulated. This is why, supported by grants of the Federal Ministry for Education and Research (BMBF), further clarification of legal frameworks, and self-regulatory activities are currently taking place. The Telematic Platform for Medical Research Networks (TMF), which provides services to the medical competence networks (KN), has aimed to clarify legal norms and technical standards. It strives to provide a 'generic model' for the establishment and the operation of biobanks. The project included expert reports on legal issues (Simon *et al.* 2006), technical data protection, quality assurance, and IT. Its genesis was tied to consensus-building processes, developed in workshops in which KN-members actively participated, which allowed for legal interpretation in the 'best' interest of the research community. Additionally, checklists and standard texts for research contracts as well as for informed consent forms will be provided (TMF 2005, 2006). Should these generic standards proposed by TMF be viable, as tested in some pilot cases, they may even be indirectly enforced: Compliance may constitute a future funding prerequisite by the Federal Ministry (BMBF). This would mean that the rules developed in self-regulatory processes may be transposed into national funding guidelines or even some day adopted by national legislation.

Concerning further governance of biobanks, some insiders discuss a central regulatory agency for the registering and licensing of biobanks, and for monitoring of their activities by audits and inspections. While some researchers would regard this as bureaucratic over-regulation, others support the idea as helpful to provide oversight, thus creating trust and public acceptance, and as bypassing some problems generated by federalism.[11] Another regulatory proposal would be to sanction unauthorized access to samples and data as an offence in criminal law (TAB 2007). Some regard this as an alternative to the elaborate and costly efforts of double-codification of samples and data ('pseudonymization'). Protection and containment of biobanks against the sequestration of samples and data by state authorities remains a thorny issue (Simon *et al.* 2006: 150). The Federal Ministry of the Interior signalled in internal consultative meetings that it would never accept a statutory prohibition of such use ('research confidentiality'); nonetheless, the TAB report (2007) reiterates that such an exemption would enable samples and data to be retained within the research environment. Another concept is to create a clearing house, initially called 'BioGEMA', which would enable access and compensation schemes for old collections of samples and data. All these proposals circulate within the biobank research community; they do not form part of public discussions.

Harmonization efforts at the international level

Several supranational initiatives endeavour to harmonize legal and technical frameworks to enable international collaboration and sharing of data. KORA-gen is an active member of the Public Population Project in Genomics (P3G), a Canadian initiative dedicated to developing European and global standards for comparing and merging results from population genomic studies, and for facilitating data management (www.p3gconsortium.org) (see Gottweis and Petersen, Chapter 1). In respective OECD working groups, as yet no German experts have been participating. An emerging international epistemic community[12], with bioethicists and lawyers as spokespeople, aims to provide 'international', 'ethical' standards for genetic biobanking. Participation in workshops and publishing strategies in renowned international journals serve as agenda and guidance-setting processes. Once guidelines are developed at the supranational level, these could possibly be 'reintroduced' to the national level, and be effective as a soft regulatory framework which may gain momentum as being 'mandatory', if the country does not want to lose track in international collaboration and competition. However, the interaction between the 'global' and the 'local' level is a complex process with potential for contradictions, failures in synchronization, and clashes.

To conclude, the emerging governance regime for biobanking in Germany comprises federal funding decisions, expert advisory commissions, pre-existing monitoring bodies, and professional self-regulatory activities. In the biobanking arena, the parliament and civil society forces are thus far mostly absent, and so is public deliberation on the value and implications of population-based biobanking. In the last section, I turn to governance *through* biobanks, its effects on medicine and the health system.

Biobanks and the future of medicine: objectives, opinions, visions

Narratives to justify the establishment of biobanks and to motivate enrolment of participants resemble those in other countries. Citizens are called upon to invest in the future of medicine: 'Tomorrow's medicine will be different. Quick diagnosis, individual therapies, efficient prevention. With a small blood sample and half an hour's time you can contribute to enable this new medicine in Schleswig-Holstein. PopGen – I help, because I know how important it is' (PopGen leaflet). An overall concept is 'individualized medicine' which means the promise of medication adapted to stratified genotypes. This may prove useful for an improved selection of therapies rather than for the development of individualized therapy regimes (cf. Schmedders *et al.* 2003). Another concept is risk prevention, which means that knowledge about a 'heightened individual risk' profile may be an incentive for personal changes in behaviour and life style. The underlying assumption is that one should take more responsibility for one's own health and that a change of 'genetic fate' is possible.

Doubtlessly, population-based genetic epidemiology will increase the global body of knowledge, and will alter aetiology and classifications of diseases. However, even the protagonists are sceptical whether this will actually provide new therapeutics, or just more genetic tests and biomarkers with uncertain validity and reliability, which may subject individuals to a life 'at risk' and cast a shadow over their future. Whether prevention of disease onset – an idea sometimes referred to as 'molecular prophylaxis' – will be possible by preventive medication, is also highly uncertain. This is not least because nobody knows whether industry would be willing to invest, and whether health insurances would be willing to pay for the lifelong administration of prophylactic drugs to 'healthy patients'. The vagueness and contingencies concerning the medical outcomes of genomics are expressed in the metaphor of the moon landing: 'You need big objectives to do big things: The idea in the late fifties that everybody may fly to the moon, was surely illusionary . . . The goal is in fact to use the genetic information of humans for a better medicine' (Schreiber, PopGen, in: Schindele 2004). The contrast between the official publicity to enrol participants which gives a very optimistic outlook, and the personal caution even among those researchers who initiated biobank projects, is striking: 'we already know that most variation in human disease is due to diet and lifestyle factors, and quantifying how the risks vary with one's genetic make-up usually won't change the solution: encouraging healthier lifestyles. The enormous investment in genomic medicine might divert resources from prevention' (Wichmann *et al.* [KORA-gen] 2005: 586). The gap between diagnostics and prevention may thus expand. Whether a new class of therapeutic products will be the output remains uncertain: 'The successful effects of biobanks, for instance in the area of pharmacogenetics, must bear comparison with other important inventions in medicine. Nowadays, everybody can experience the effects of antibiotics or X-rays and appreciate them. Similar influential developments from genetic technologies have as yet not been noted' (Krawczak, PopGen, in: TMF 2005: 212).

However, there is hardly a public sphere to honestly address these uncertainties within the globalized, competitive, scientific, and economic race, which is part and parcel of modern biomedicine and its entrepreneurial, market-driven R&D model of health innovation (cf. Nightingale and Martin 2004). What amount of financial resources, both public and private, should be allocated to the (post)genomics project, according to which priorities, and how this will shape national and international healthcare systems, remain questions which deserve more democratic deliberation (Williams 2005: 64; Merikangas and Risch 2003). Within this debate, it may be important to consider Neil Holtzman's warning: 'Exaggerating the importance of genetic factors as determinants of health stops people thinking about the need to clean up the environment and tackle socioeconomic inequity' (Burn *et al.* 2001).

The future of biobanks

Biobanks are cost-intensive endeavours. If they should be of use, they must be considered as a permanent infrastructure project – similar to streets, telecommunication, or archives – to be continuously administered. German researchers emphasize this point by distinguishing between an 'active' and a 'passive' concept of biobanks: Biobanks just storing DNA and data are called '*Sparstrumpf*', which is money saved under the mattress. In contrast, 'active' biobanks, creating surplus value, require investment. Not incidentally, the pervasive 'bank' metaphor has gained supremacy over the notion of 'biorepositories' or 'biolibraries'. Obviously, biobanks can only be useful if they are used. However, any use will also have effects on the biobank itself – it will require permanent remodelling and constant reinterpretation of the samples and data collected. Whether biobanks will be frozen DNA tombs and data cemeteries or a useful infrastructure for a post-genomic era remains unclear. There seems to be an in-built institutional impetus for self-preservation and growth: Thus, any successful establishment of such a cost-intensive infrastructure may warrant new research grants and investments to make efficient use of the 'precious resource'. Will politicians follow the Sirens' call, and will private companies further step in? Continuing funding requirements may seduce researchers to exaggerate the value of biobanks, to oversell their potential. This may be a hindrance for a more rational and informed public debate about the promises, potentials and dangers of genetic biobanking. Will the inflated economies of hope and expectation suffer the same fate as the bursting biotech bubbles at the stock markets?

To some extent, critical perceptions of biobanks as instruments of panoptical surveillance or micropolitical control strategies may be seen as mirroring the genohype and thus be apologetic in this respect. At least, concerns about totalizing biopolitical surveillance and 'transparent' citizens seem not to be well-founded in Germany where size and scale of biobanks cover only a very small percentage of the population. However, once larger-scale data and sample collections are created, they may generate new desires or even new 'moral obligations' to disclose and make use of their content, be it on the personal or on the state level; they may be utilized for purposes which were previously unanticipated.

Conclusion

In explaining why as yet there is no 'Biobank Germany', the politics of local biobanks have been elucidated. Constitutive factors identified are the strong federalism of the German polity, research funding practices, structures of the health system, the political conflict culture, legal frameworks, and constitutional norms. The understatement in the tone – superlatives are avoided – and the 'localized' naming and framing of German biobanking activities have several advantages for the actors: It has facilitated governance that is largely left to professional self-regulation. It allowed for depoliticization, the taming of potential conflict, thus obviating possibly restrictive statutory measures, or demands for more transparency, accountability, and public participation. By the same token, it has enabled national and international consensus building processes within an emerging epistemic community, driven by professional self-interests and the advancement of new research paradigms. Biobanks are thus cultural mediums through which the proliferation of geneticized medicine will take course.

Restraints posed by the 'local' framing of biobanks may be shown by the fact that these biobanks have so far not been very appealing, neither for non-profit foundations nor for pharmaceutical companies or venture capital. The 'lack of hype', which succeeded in suppressing the potential for public opposition, may be detrimental to the generation of finance. The major public biobanks have so far not sought private funding, but exhibit a strong commitment to a public mission. However, short federal funding periods and the need to provide for future self-financing may exert pressure on finding partners from the private sector to maintain the infrastructure. Whether this will influence the biobank's design and the direction of research performed remains to be seen. The localized approach may also possibly effect an 'underuse' of the databases. So far, there is some competition between local biobanks, and hardly any coordinated action, which points to problems of collective action. The fact that there are as yet no homogeneous standards may also hamper the scientific comparability of results.

To conclude, in Germany as yet there is no 'national' biobank, but two population-based genetic biobanks, which can possibly 'compete' if not in quantity then maybe in quality with major approaches in other countries. Time will show whether several smaller, disease-specific biobanks which enable specific phenotyping, or whether single, large 'catch-all' approaches will yield convincing scientific results. Heterogeneous national approaches may even result in mutual checks and balances. However, the divergences may also make it impossible to replicate results attained in other research designs. In fact, there are already many cooperation processes going on, which do not operate at the trans-national but on the trans-regional level (see KORA/MONICA). So, in lacking a 'national biobank', Germany may, on closer inspection, not be an exception to the global biobank landscape, but very much the rule.

Acknowledgements

This chapter draws on an interim report on Biobanks in Germany by Nikolaus Zacherl for the ELSA Biobanks project, funded by the GEN-AU (Genome Research in Austria) programme of the Austrian Federal Ministry for Education, Science, and Culture, for which I express my sincere gratitude. It is also based on interviews and conversations with researchers from German biobank projects, members of the National Ethics Council, lawyers, and members of the Bundestag, and on participatory observation at several workshops, and public conferences of the National Ethics Council, the Office of Technology Assessment (TAB) of the Bundestag, the TMF and the NGNF. I am indebted to all my conversation partners for frankly sharing their views and assessments. The views expressed in this article only reflect those of the author. The responsibility for any remaining errors is solely mine. All German quotes were translated by the author.

Notes

1 The title alludes to René Magritte's 1928–9 painting of a pipe entitled *Ceci n'est pas une pipe*. It thus evokes the dimensions of representation and semantics concerning biobanks.

2 The LURIC-Study (LUdwigshafen RIsk and Cardiovascular Health) for instance is based on collaboration between the Ludwigshafen General Hospital, the pharmaceutical company Aventis, and scientists at the Universities of Freiburg, Ulm and Graz.

3 Genetic heterogeneity seems to be anathema for German researchers, at least in its public representation. Nonetheless, the sensitive issue of genetic diversity is subtly addressed: One of the inclusion criteria for PopGen's control cohort is that the research subject's family has been living in Germany for the last three generations (personal communication, M. Krawczak, 12 December 2005). As migration from Southern and extra-European territories occurred mostly in the last four decades, thereby an exclusion process is taking place, although the term 'Caucasian' sample is avoided. However, in a recent study, genetic diversity in German and European populations was hypothesized as causal for the fact that only a small fraction of significant genetic association results could be replicated (Lamina *et al.* 2005).

4 Information provided by Huberta von Eberstein, PopGen, by telephone, 8 June 2006.

5 Krawczak, personal communication, 12 December 2005.

6 In this respect, developments in the USA and in France may be of crucial importance. See for the US: Secretary's Advisory Committee on Genetics, Health, and Society: Policy Issues Associated with Undertaking a Large US Population Cohort Project on Genes, Environment, and Disease. Draft report, May 2006, available at www4.od.nih.gov/oba/SACGHS/public_comments.htm (accessed 3 July 2006).

7 In contrast to Anglo-Saxon common law jurisdictions, most continental European civil law jurisdictions have specific civil code provisions that protect an individual's personal data, image, privacy, personality and publicity. In Germany, 'personality rights' are protected under the German Civil Code and in constitutional law. These rights include 'the control of bodily and intellectual faculties' and all fundamental civil liberties. The 'general right of personality' includes the right to privacy, prevents harm to life, and protects physical integrity, health, and liberty. It encompasses a right to informational self-determination, which *inter alia* imposes an obligation on the state to organize the collection, storage, and transmission of information in a way that respects the individual's personal autonomy. Furthermore, personality rights also regulate some relationships between individuals. As a consequence, financial compensations for violations of personality rights are possible (cf. Le Bris and Knoppers 1997: 428f.). It should, however, be noted that the juridical category of 'personality rights' allows different cognitive and legal conceptions of issues pertaining to bodily material and information which go far beyond 'property rights'.

8 Withdrawal in Germany is constitutionally backed as an inalienable right deriving from the right on self-determination on the use of personal *data* and from the legal concept of continuing *personality rights* – representing the will and integrity of the person – which are tied to the *material* bodily sample (Halàsz 2004; Simon *et al.* 2006).

9 This proposal is particularly emphasized in the Joint Declaration on biobanks or 'biothèques' by the German and the French National Ethics Councils (see NEC 2004, 101–2).

10 The German term is '*Forschungsgeheimnis*', which in the literal translation means 'research secrecy' or could be translated as 'research privilege', shielding researchers from undue encroachment by state authorities, similar to special protection of confidentiality between medical doctors and their patients, lawyers and clients, and also journalists and their informants.

11 Interview with Sebastian Semler, managing director of TMF, 27 April 2006.
12 For the concept see Haas (1992). The biobanking epistemic community is composed of clinical researchers and theoretical molecular scientists, epidemiologists, bioinformaticists, bioethicists, lawyers, and members of some other disciplines (see Knoppers and Chadwick 2005; Cambon-Thomsen 2004; Chadwick and Berg 2001).

References

BMBF (2002) *Kompetenznetze in der Medizin. Forschung für den Menschen*, available online at www.bmbf.de/pub/kompetenznetze_in_der_medizin.pdf (accessed 13 June 2006).

BMBF (2003) *Das Nationale Genomforschungsnetz. Krankheitsbekämpfung durch Genomforschung*, available online at www.bmbf.de/pub/das_nationale_genomforschungsnetz.pdf) (accessed 13 June 2006).

BMGS (Federal Ministry of Health and Social Welfare) (2004) *Gesetz über genetische Untersuchungen bei Menschen* (Gendiagnostikgesetz – GenDG), Version 15 December 2004 (unpublished draft act).

Burn, J., Duff, G. and Holtzman, N. A. (2001) 'Three views of genetics: the enthusiast, the visionary, and the sceptic', *British Medical Journal*, 322: 1016.

Cambon-Thomsen, A. (2004) 'The social and ethical issues of post-genomic human biobanks in ethics', *Nature Reviews Genetics*, 5: 866–73.

Chadwick, R. and K. Berg (2001) 'Solidarity and equity: new ethical frameworks for genetic databases', *Nature Reviews Genetics*, 1 (2): 318–21.

Goebel, J. (2006) *Persönlichkeitsrechte versus Forschungsvisionen – Rechtssicherheit für die langfristige Nutzung der Materialien*, available online at www.tmf-ev.de/site/DE/int/news/news_extern/Symposium_BMB_04–2006/Doc/2a_Goebel_BMB-Symposium_27-04-2006.pdf (accessed 13 June 2006).

Haas, P. M. (1992) 'Introduction: epistemic communities and international policy coordination', *International Organization*, 46 (1): 1–35.

Habermas, J. (1986) 'Eine Art Schadensabwicklung. Die apologetischen Tendenzen in der deutschen Zeitgeschichtsschreibung', *Die Zeit* (10 July): 62–76.

Halàsz, Christian (2004) *Das Recht auf bio-materielle Selbstbestimmung: Grenzen und Möglichkeiten der Weiterverwendung von Körpersubstanzen*. Berlin: Springer.

Halliday, J. L., Collins, V. R., Aitken, M. A., Martin, P. M. and Olsson, C. A. (2004) 'Genetics and public health – evolution, or revolution?', *Journal of Epidemiology and Community Health*, 58 (11): 894–9.

Holle, R., Happich, M., Löwel, H. and Wichmann, H. E. (2005) 'KORA – a research platform for population based health research', *Gesundheitswesen*, 67 (suppl. 1): 19–25.

Holtzman, N. A. and Marteau, T. (2000) 'Will genetics revolutionize medicine?', *The New England Journal of Medicine*, 343 (2): 141–4.

Jasanoff, S. (2005) *Designs on Nature: Science and Democracy in Europe and the United States*. Princeton, NJ, and Oxford: Princeton University Press.

Knoppers, B. M. (2004) 'Biobanks: simplifying consent', *Nature Reviews Genetics*, 5: 485

Knoppers B. M. and Chadwick, R: (2005) 'Human genetic research: emerging trends in ethics', *Nature Review Genetics*, 6 (1): 75–9.

Krawczak, M., Nikolaus, S., von Eberstein, H., Croucher, J. P., El Mokhtari, N. E. *et al.* (2006) 'PopGen: population-based recruitment of patients and controls for the analysis of complex genotype-phenotype relationships', *Community Genetics*, 9: 55–61.

Lamina, C., Steffens, M., Mueller, J., Lohmussaar, E., Meitinger, T. *et al.* (2005) 'Genetic diversity in German and European populations: looking for substructures and genetic patterns', *Gesundheitswesen*, 67: 127–31.

Le Bris, S. and Knoppers, B. (1997) 'International and comparative concepts of privacy', in M. Rothstein (ed.), *Genetic Secrets. Protecting Privacy and Confidentiality in the Genetic Era* (pp. 418–48). New Haven, CT, London: Yale University Press.

Lippmann, A. (1991) 'The geneticization of health and illness: implications for social practice', *Endocrinology*, 29 (1–2): 85–90.

Merikangas, K. R. and Risch, N. (2003) 'Genomic priorities and public health', *Science*, 302: 599–601.

NEC (National Ethics Council) (2004) *Biobanks for Research, Opinion* (17 March 2004), available online at www.ethikrat.org/_english/publications/opinions.html (accessed 13 June 2006).

Nightingale, P. and Martin, P. (2004) 'The myth of the biotech revolution', *Trends in Biotechnology*, 22: 564–9.

Roos, D. S. (2001) 'Bioinformatics – trying to swim in a sea of data', *Science*, 291: 1260–1.

SACGHS (Secretary's Advisory Committee on Genetics, Health, and Society) (2006) 'Policy issues associated with undertaking a large U.S. population cohort project on genes, environment, and disease.' Draft report, May 2006, available online at www4.od.nih.gov/oba/SACGHS/public_comments.htm (accessed 3 July 2006).

Schindele, E. (2004) 'Datenspende Blut. Wie Biobanken Patientenmaterial vermarkten' (radio report, *Deutschlandfunk*, 21 November 2004), available online at www.dradio.de/dlf/sendungen/wib/323772/?drucken (accessed 13 June 2006).

Schmedders, M., van Aken, J., Feuerstein, G. and Kollek, R. *et al.* (2003) 'Individualized pharmacogenetic therapy: a critical analysis', *Community Genetics*, 6: 114–19.

Schneider, I. (2002a) 'Biobanken: Körpermaterial und Gendaten im Spannungsfeld von Gemeinwohl und privater Aneignung', *Zeitschrift für Biopolitik*, 1 (3): 31–8.

Schneider, I. (2002b) 'Ausverkauf der Gene? Gewebe- und Gendatenbanken zwischen Persönlichkeitsschutz und Kommerzialisierung', in G. Arnim, V. Deile and F. J. Hutter (eds), *Jahrbuch Menschenrechte 2003* (pp. 130–44). Frankfurt/Main: Suhrkamp.

Schreiber, S. (2003) 'Biobanken und Populationsgenetik im Deutschen Humangenomprojekt', Nationaler Ethikrat (ed.), *Biobanken. Chance für den wissenschaftlichen Fortschritt oder Ausverkauf der "Ressource" Mensch?* (pp. 25–38, 93–111). Dokumentation der Jahrestagung des Nationalen Ethikrates 2002, Berlin.

Simon, J. W., Paslack, R. and Robienski, J. (2006) 'Generisches Konzept für den Aufbau und Betrieb von Biomaterialbanken. Projekt der Telematikplattform für Medizinische Forschungsnetze (TMF) e.V., Teilprojekt 1', *Rechtliche Fragen.* Gutachten.

TAB (Office of Technology Assessment at the German Parliament) (2007): *Biobanks for Human Medical Research and Application*, available online at www.tab.fzk.de/en/projekt/skizze/biobanken.htm; published as Ch. Revermann and A. Sauter (2007) *Biobanken als Ressource der Humanmedizin. Bedeutung, Nutzen, Rahmenbedingungen*, Berlin: Sigma.

TMF (Telematikplattform für Medizinische Forschungsnetze) (2005) 'Bestandsaufnahme und Charakterisierung von Biobanken – Systematisierung, wissenschaftliche Bewertung, Finanzierungsmodelle und Konzepte zu Datenschutz und Patienteneinwilligung'. Report commissioned by the German Bundestag for the TAB, Berlin.

TMF (2006) TMF-Symposium 'Biomaterialbanken: Zwischen Überregelung und Wildwuchs', public conference, Berlin, 27 April 2006), available online at www.tmf-ev.de/site/DE/int/news/news_extern/Symposium_BMB_04–2006/c_BMB-Symp.php.

Wagenmann, U. (2005) *Charakterisierung von Biobanken im Hinblick auf Gesundheitspolitik und Medizin.* Report commissioned by the German Bundestag for the TAB, Berlin.

Wichmann, H. E., Gieger, C. and Illig, T. (2005) 'KORA-gen – resource for population genetics, controls and a broad spectrum of disease phenotypes, *Gesundheitswesen*, 67 (suppl. 1): 26–30.

Williams, G. (2005) 'Bioethics and large-scale biobanking: individualistic ethics and collective projects', *Genomics, Society and Policy*, 1 (2): 50–66.

Winickoff, D. E. (2003) 'Governing population genomics: law, bioethics, and biopolitics in three case studies', *Jurimetrics*, 43: 187–228.

Winickoff, D. E. and Neumann, Larissa B. (2005) 'Towards a social contract for genomics: property and the public in the "Biotrust" model', *Genomics, Society and Policy*, 1 (3): 8–21.

7 Governing DNA

Prospects and problems in the proposed large-scale United States population cohort study

Amy Fletcher

Introduction

In the United States, biomedical topics such as stem cell research and pre-implantation genetic diagnosis generate media coverage and controversy. Such issues can be placed within the highly salient frame of the abortion issue, ensuring that legislators and a large segment of the public will engage with them. However, despite the establishment of national biobanks in many European countries, the United States only recently initiated public consideration of whether or not it needs a large-scale, prospective national cohort for the study of the relationship between genes, disease and the environment. This chapter analyzes this initial discussion (2004–6), with specific reference to the political and institutional variables that impede the construction of an effective governance regime for a national biobank in the United States.

In May 2004, the National Institutes of Health (NIH) released a formal request for information, seeking 'advice on approaches to developing a large-scale US study of genetic and environmental influences on common diseases'(NIH 2004a: 1). A draft report for public comment on the policy issues associated with a large-scale US population cohort on genes, environment and disease followed in May 2006, from the Secretary of Health's Advisory Committee on Genetics, Health and Society (SACGHS). In September 2006, the National Human Genome Research Institute (NHGRI), an agency within the National Institutes of Health, awarded US$2 million to the Genetics and Public Policy Center (GPPC) in Washington, DC, to conduct a pilot study of 'the public's hopes and concerns regarding such large-scale studies to help find the underlying causes of illness' (GPPC 2006). A preliminary analysis of the GPPC pilot consultation will not be complete until late 2008, and will then be 'incorporated into the design of the longitudinal cohort study, its full-scale public consultation component, and other population-based studies, should they be determined to be feasible and should they be funded within the next few years' (SACGHS 2006: 50). It is unlikely that a final decision to proceed with a large-scale national cohort will be made in the near future. However, the initial deliberations regarding a large-scale cohort in the United States provide a platform to consider the theory and practice of biomedical governance in the contemporary American context. This chapter therefore asks: what governance challenges are raised by the NIH proposal in relation to a large-scale prospective cohort in the United States?

Governance is defined herein as both governmental actions and 'other processes, formal and informal, that communities employ to decide what is in their common interest, and how to act collectively' (Speth and Haas 2006: 3). In the US case, both the scale and demographic diversity of the country mean that governance challenges include appropriately delimiting the scope of the community to be consulted, defending the 'common interest' in health sector reform, and conducting meaningful public consultation in a highly diverse country of 300 million people. In order to build and sustain public trust in such a large-scale project, the Department of Health and Human Services (DHHS) will also need to balance the role of disease advocacy groups against the general public interest, and support the long-term development of 'personalized medicine' while (in cooperation with the Food and Drug Administration) constraining fantastic commercial claims about the efficacy and availability of drugs targeted to individual genetic profiles. Catherine Lyall argues 'we have seen a shift from scientific/economic governance to more social forms of governance, but different discourses must still co-exist: science-based, ethics-based, competitiveness focused, risk v. benefit approaches' (Lyall 2005). This observation holds true in the US case, where the multitude of potential political interests combined with multiple channels of citizen entry into the formal decision-making process maximizes the number of contending discourses and strategies.

The next two sections discuss the SACGHS draft report on a prospective national cohort study, in the context of both existing public and private biobanks, and the unique characteristics of the American healthcare system. Section four focuses on the use of the personalized medicine frame, by both commercial and federal advocates of such a study, to promote and justify a major public investment in a prospective national cohort. The fifth section analyzes the role of advocacy organizations and community-based biobanks in the health sector. It focuses on how the need to consult organized health interest groups in the implementation of a national cohort risks distorting any common public interest in the study of genes and disease in favor of the specific interests of the most visible and well-funded organizations. The chapter concludes by linking both of these aforementioned issues to the need to establish public trust in the Public Health Service (PHS) of sufficient depth and breadth to sustain a national biobank across the necessary time span of at least two decades, and across a highly pluralistic and ethnically diverse country.

A wealth of biobanks but no national cohort

Biobanks are 'organized collections of biological samples and the data associated with them', and 'come in many different forms, according to the types of samples that are stored and the domain in which they are collected' (Cambon-Thomsen 2004: 866). It is alongside increasing media coverage of the Human Genome Project in the late 1990s that the concept of national prospective cohorts (also known as national biobanks) emerges on the policy agenda in many European and Asian countries. SACGHS, for example, in preparing its draft report for public comment, studied large-scale cohorts either established, or under discussion, in the United Kingdom, Iceland, Estonia, Germany Canada, Taiwan, and Japan. In the United States, many public and private biobanks combine DNA samples with clinical information for health-related research. Numerous stakeholders that want to manage the value – whether social or economic – of genetic information also seek either to establish new biobanks or to create policy networks that can influence legislation in related areas such as genetic property rights and genetic discrimination.

The explosion of biomedical/genomic research since the completion of the Human Genome Project (HGP) indicates the priority given to translating expensive twentieth-century basic research on genomics into applied health outcomes. For example, in January 2006, the National Cancer Institute (NCI) and the National Human Genome Research Institute (NHGRI), two affiliates of the NIH, launched the Cancer Genome Atlas (TCGA) pilot project. The three-year pilot, funded at US$100 million, is the initial stage in a long-term effort to develop a Cancer Genome Atlas that 'will describe the genetic "fingerprints" of specific cancer types' (NCI 2006a) and contribute to the development of effective treatments for all types of cancer. Tumor samples and clinical information will be gathered voluntarily from cancer patients undergoing treatment. The Department of Veterans Affairs (VA) also has considered creating a gene bank that would link DNA donated by up to seven million veterans (and their family members) with its extensive collection of medical records. This proposed biobank 'is widely supported inside and outside the VA', and in the absence of a national cohort implemented by the NIH, 'would represent the first massive U.S. gene banking effort' (Couzin 2005: 684).

The National Cancer Institute also launched the Consortium of Cohorts in 2000 in order to 'address the need for large-scale collaborations for study of gene-gene and gene-environment interactions in the etiology of cancer' (NCI 2006b). The consortium combines ten existing large cohorts to create a research pool of 'nearly 800,000 research participants with available biospecimens' (NCI 2006b). In addition, the Epidemiology and Genetics Research Program (EGRP) at NIH supports several ongoing large-scale cohorts such as the Black Women's Cohort (Follow-Up Study for Causes of Illness in Black Women), Cancer in American Natives (A Prospective

Study of Alaska Natives and American Indians), and the Southern Community Cohort Study.

In addition to such large-scale federal efforts, the US biobank sector includes academic, private, and non-profit biobanks. Academic collections, established under the direction of a scientist or research team affiliated with a university, generally have specific disease or cohort research goals and a community outreach function. For instance, the Center for Acadiana Genetics and Hereditary healthcare at Louisiana State University focuses on local descendants of the Acadians who in settled in South Louisiana in the nineteenth century. This biobank combines DNA samples, clinical data, and the detailed genealogical records kept by Acadian families and churches. Another example, the Kosrae Project at Rockefeller University, focuses on the heritability of disease via extensive genomic and clinical mapping of the 4,000 residents of Kosrae Island in the West Pacific. Non-profit biobanks often belong to advocacy organizations that focus on specific diseases. Examples of non-profit biobanks include the Alpha-1 Foundation DNA and Tissue Bank, the Angioma Alliance Tissue/DNA Bank and Patient Registry, the National Psoriasis Biobank, and the Autism Genetic Resources Exchange (AGRE). The non-profit biobank sector also includes large-scale cohort studies conducted by major public clinics. The Marshfield Clinic Personalized Medicine Research Project (PMRP), for example, seeks to 'translate genetic data into specific knowledge about disease that is clinically relevant and will enhance patient care' (Marshfield Clinic 2006). Finally, private biobanks exist in the United States, such as the one owned by the Church of Jesus Christ of Latter-day Saints (LDS) in Utah (Rosen 2003). LDS also launched the Molecular Genealogy Research Project (which is now funded and managed by the non-profit Sorenson Foundation).

The biobank/cohort examples cited herein illustrate both the complexity and breadth of the issue, as well as the acceleration in genomic research since the completion of the Human Genome Project. Existing stakeholders with collections that are several decades old, such as the Marshfield Clinic in Wisconsin, and the Framingham Heart Study in Massachusetts, want to repackage and update their data in a form appropriate for genomic research and drug development. New stakeholders want to establish new biobanks focused on under-researched cohorts (particularly in the field of rare diseases) or to collaborate in the creation of alliances that can leverage access to several collections at once. Despite the vast amount of biobank activity in the United States, however, the country does not have a large-scale prospective national cohort similar to, for instance, the UK Biobank. Such a public project would enroll at least 500,000 people, with associated medical data and tissue samples, in order to study the emergence of diseases – and the relationship between genes, environmental exposures, and disease – over decades. The NIH argues that 'rigorous and unbiased conclusions about disease etiology and population impact will require prospective population-representative designs' (NIH 2004: 1). Similarly, SACGHS, in its 2006 draft report, uses the term 'project' in reference to 'the longitudinal collection and storage of data and biological specimens from large numbers of people for the research use of multiple investigators and investigative teams' (SACGHS 2006: 4).

The prospect of an American national cohort dovetails with recent academic attention to the challenges of governance in the United States. Putnam's famous analysis of the decline in American social capital since the 1950s (Putman 2000) and scholarly attention to the decline in trust in government (see, for instance, Nye *et al.* 1997) are key works in a research agenda focused on 'knowing more about the role of civic engagement as a central component of a vital American democracy' (Cooper *et al.* 2006: 76). The following section analyzes potential obstacles to effective biobank governance in the United States, including the complexity of the healthcare system and funding constraints for long-term non-defense related health research.

Governing DNA: Biobanks and the US healthcare system

With respect to deliberations about national cohorts/biobanks, certain policy issues routinely emerge in most countries. Common concerns include questions regarding individual privacy rights, the ownership of genetic information, the nature and meaning of 'informed consent', and the most effective and ethical way to enroll citizens in a project of potentially open-ended duration. Factors that further complicate the US case include the lack of universal healthcare (raising the possibility that the uninsured will remain outside the national cohort) and the related lack of a uniform record-keeping system.

In addition, the institutional combination of federalism, in which fifty states and the national government vie for political authority in the health sector, with pluralism, which encourages the formation and participation of interest groups, ensures regulatory complexity. The separate (and overlapping) powers of the executive, legislative and judicial branches of government further mean that many biomedical issues – particularly those related to privacy and consent – will be decided in the courts via an adversarial judicial process. Financially, an analysis by the German Association of Research-Based Pharmaceutical Companies (VFA) indicates that from 2001 to 2004, global pharmaceutical sales reached a total value of US$550 billion. Almost half of these worldwide sales are generated in the United States (VFA 2006). The size and global economic clout of the United States pharmaceutical industry ensures that any proposal to establish a large-scale prospective cohort will activate powerful stakeholders to participate directly in the delineation of new regulations and research goals. SACGHS recognizes these challenges in its draft report:

> A large population study in the United States could be difficult because it will put pressure on the American healthcare system, which is characterized by uncoordinated, decentralized private and public institutions.

> Thus, given the current fragmented state of healthcare in this country, some members of the scientific community are asking whether a truly coherent cohort study can be designed, data collected and analyzed, and benefit returned to the participants and others at a reasonable cost.
> (SACGHS 2006: 14).

Consistent funding for a major health project such as a national cohort could also be problematic. In September 2006, NIH noted that 'although funding for such an endeavor has not been identified, carefully outlining and considering the goals and key design aspects of such studies was deemed of high scientific importance' (NIH 2006: 1). Though supportive of the deliberations on a national cohort, the American Association of Medical Colleges (AAMC) is more explicit:

> in fact, there are severe funding constraints across all Public Health Service (PHS) programs and all Federal discretionary programs with the exception of Defense and Homeland Security . . . The AAMC recommends that the final report make it clear that the resource implications for the proposed study may be far more acute in the current and foreseeable funding environment than currently stated.
> (AAMC 2006: 2).

The history of the National Children's Study (NCS) reflects the scenario outlined by the AAMC above, supporting the argument that a major obstacle to the long-term viability of a national cohort will be political/fiscal uncertainty. The NCS mission is to 'examine the effects of environmental influences [including gene–environment interactions] on the health and development of more than 100,000 children across the United States, following them from birth to age 21' (NCS 2006). Despite a six-year project planning effort by the NCS, President George W. Bush's proposed budget for fiscal year 2007 instructed the NCS to cease activity as of 30 September 2006; however, both the House and Senate Appropriations Committees indicated an intention to continue its financial support. In October 2006, the NCS website noted that resolution of the Study's future funding status would not happen until after the November 2006 Congressional elections (NCS 2006). This example illustrates a major political challenge to launching and maintaining a national cohort project whose health benefits might not be realized for decades, if at all, and whose participant uptake would be need to be approximately five to ten times larger than the Children's Study.

The politics of personalized medicine

Personalized medicine is a nascent industry model emerging in the pharmaceutical industry. American proponents of a national cohort argue that personalized medicine could 'result from understanding the variation in DNA that makes humans different from one another in their susceptibility to disease, their physiological, mental and emotional responses to physical, behavioral, and social environmental exposures, and their response to medicines' (SACGHS 2006: 15). The lure of personalized medicine anchors biobanks as the key institution for the storage, use and exchange of genetic information. Individual DNA samples and medical histories are the deposits upon which the economic system of personalized medicine depends, though it is primarily when this information is pooled into large-scale cohorts that financial growth and pharmaceutical innovation occurs. As S. Hall notes 'the defining characteristic of every form of personalized medicine is its biomarker, a kind of biological fingerprint that distinguishes a subset of the patient population' (2003: 66). Consequently, 'interest' accrues to both society and individuals if and when projects based on biobank data discover biomarkers that can be translated into profitable genetic tests and personalized drugs. Biobanks gain tangible social and economic value (in addition to basic scientific value) when the data they contain potentially translate into progress toward personalized medicine.

The personalized medicine construct presents rational drug design (based on biomarkers and other genomic techniques) as a revolution in healthcare – a dramatic departure from existing practice comparable to previous advances such as the germ theory of disease or vaccination campaigns. For example, Eric Lander, Director of the Human Genome Center at the Whitehead Institute (MIT), believes 'people looking back 50 years from now will consider medicine a barbaric, random process. If the promise of genomics is fulfilled, it will transform the lives of everyone' (quoted in Fischer 2001). A report from Price Waterhouse Coopers (PWC) also links pharmacogenomics to 'revolutionary change' such as antibiotics and vaccines, and argues that pharmacogenomics 'promises to usher in an era of individualized patient care or personalized medicine' (PWC 2005: 1).

The lure of personalized medicine links public and private incentives to support a national cohort via promises of substantial returns on public investment in the form of more effective and personalized treatments and medication; indeed, the personalized medicine slogan of 'the right drug at the right dose to the right person at the right time' dominates discussion of genomics at the federal level. In November 2005, the National Human Genome Research Institute met to discuss the challenge of how to translate the explosion in human genomic knowledge into improved healthcare. It established a major goal of enhancing 'health care in the USA through the integration of genomic medicine into mainstream health practice', with an emphasis on the design of more effective drugs, individualized treatments and predictive genetic tests (NHGRI 2005). The Cancer Genome Atlas (TCGA) project leaders also frame their programme within the context of personalized medicine. NIH Director Elias A. Zerhouni states that these 'new insights into the biological basis of cancer [will] lead to new tests to detect cancer in its early, most treatable stages; new therapies to target cancer at its most vulnerable points; and, ultimately, new strategies to prevent cancer' (NIH 2005).

The prospect of personalized medicine also increases the economic value of genetic information; as *The New York Times* notes 'estimates say that one human body can bring in anywhere from $10,000 to nearly $150,000. That's nothing compared with DNA – just one gene can be worth millions' (Skloot 2006). New stakeholder groups and policy networks in the United States health sector want to capitalize on this increased value. The Personalized Medicine Coalition (PMC) – launched in 2004 – defines personalized medicine as 'the use of new methods of molecular analysis to better manage a patient's disease or predisposition towards a disease' (PMC 2006). It sponsors public forums on genomics and medical applications and actively seeks to influence federal legislation on genetic non-discrimination and other related priorities. Faster Cures/The Center for Accelerating Medical Solutions, an initiative of the Milken Institute, is 'committed to accelerating the medical research process to find new treatments for deadly and debilitating diseases' (Faster Cures 2006). It houses BioBank Central, arguing that biobanks are essential to 'harnessing the power of both genomic and clinical data, [serving] as a critical bridge between basic and applied research, linking laboratory to patient and getting to cures faster' (Faster Cures 2006). BioBank Central, in turn, is sponsored by IBM, Affymetrix, Bioaccelerate, and Invitrogen.

Yet tying the worth of a large-scale national cohort to the promise of personalized medicine is a risky strategy. An editorial on the UK Biobank argues 'such studies are marketed to the public in a tone of high adventure appropriate for proposed climbs of Mt. Everest than mundane scientific research' (*The New Atlantis* 2003: 102). This is also true in the United States, where supporters of a national cohort risk the erosion of future public confidence in both large-scale projects and deep public investment in health if effective and safe personalized drugs do not rapidly emerge following the launch of a US Biobank. Moreover, in the United States the pharmaceutical industry 'is the only route available to develop new products from the huge and increasing public charitable and private investments in the generation of new knowledge from genomics and related fields' (Tait and Mittra 2004: 1). A successful national biobank project in the United States will thus need to balance the NIH's public education function – muting public expectations via a forthright discussion of the pace and current limits of genomic science – against its role in encouraging public investment and participation now by highlighting potential health benefits. Regulations for the personalized medicine industry must be drawn in a way that supports the pharmaceutical industry as the engine of this new sector, while constraining incentives to exaggerate the 'revolutionary' status of personalized medicine beyond what the scientific knowledge base can ethically support. Even proponents of a national cohort such as Francis Collins conclude that 'although genome-based analysis methods are rapidly permeating biomedical research, the challenge of establishing robust paths from genomic information to improved human health remains immense' (Collins *et al.* 2003: 836). Moreover, as the next section demonstrates, the influence of organized disease advocacy groups in determining which diseases receive research emphasis could further erode public confidence in the priority-setting legitimacy of the NIH.

Power to the patient: balancing interests in a large scale national cohort

Much of the contemporary research on governance presumes that when stakeholders in an issue 'are treated as potential peers, sharing the definition and management of a problem, they can mobilize resources of local knowledge and understanding which complement the generalized knowledge of scientists or official experts' (De Marchi and Ravetz 1999: 743). Pluralism in theory and in practice is fundamental to American politics, dating to the publication of *The Federalist Papers* in the late eighteenth century. The US system encourages the formation of interest groups, and anticipates competition between them for funding, visibility and access to the various branches of government. In this general context, the idea of large-scale cohort built around a model of deep public consultation fits seamlessly within traditional American norms and aspirations. In turn, such a project could reflect and reinforce new administrative models of 'citizen-centered collaborative public management' (Cooper *et al.* 2006: 76). The draft report acknowledges the crucial role of organized groups in the health system, and explicitly refers to the need to consult disease advocacy groups and community organizations at 'several points along the project timeline, in order to support the very concept of such a project and to sustain public trust and interest in its continuance' (SACGHS 2006: 21).

With specific reference to health sector government, this emphasis on extensive and multi-channel community input reflects a turn from hierarchical and expert-driven medical governance in the twentieth century to a more open and collaborative science–society–government relationship wherein 'collective problems are being solved more and more by state and non-state actors collaborating across levels' (Finger 2004: 4). Patients and their organized advocates in the United States can now use biobanks as resources for ensuring influence over research and product development. Just as individual patients gain more power by forming non-profit organizations and disease advocacy groups, these organizations are also forming alliances in order to amass 'gene power' relative to the state and industry. For example, in 2003, the Genetic Alliance, an advocacy group that seeks to leverage the collective power of organized patient groups (and their collections of associated medical records and tissue repositories) established the Genetic Alliance BioBank to exert formal control over genomic research on several diseases. Founder Sharon Terry argues:

> The BioBank is evidence of the next generation of patient advocacy. But this is only the beginning. We are managing this resource, this community, with our eye on the prize – we will positively impact health outcomes.
> (Genetic Alliance Biobank 2004)

Current members of the Genetic Alliance BioBank include PXE International, and the National Psoriasis BioBank, while the Genetic Alliance, in turn, is a member of the Personalized Medicine Coalition and the Coalition for Genetic Fairness.

The Genetic Alliance BioBank builds on an organizational model that founders Sharon and Patrick Terry created upon receiving the news that two of their children have PXE – a rare disease that affects approximately 11,000 patients in the United States and which can result in premature death. Given the small size of the market for treatments, PXE did not normally attract a lot of investment or attention from pharmaceutical companies. The Terry's decided to represent PXE patients not only via traditional financial and legislative activities, but also by creating a biobank through which the PXE community could influence research as the gatekeepers to a valuable collection of tissue samples and clinical information. Furthermore, Sharon Terry's research contributions to the search for the PXE gene resulted not only in her listing as an author on a major paper published in *Nature Genetics*, but also the assignment of the patent rights to ABCC6 (a gene related to PXE) to PXE International. Similarly, CFC International, a support group for approximately 100 patients and their families, decided in 2004 to join with several other genetic disease groups to set up a central biobank of patient records and DNA samples. The tissue specimens and medical data enabled Dr Katherine Rauen and a research team at the University of California, San Francisco, 'to find mutations in three genes, BRAF, MEK1, and MEK2, that explain 21 of the 23 CFC cases they examined' (Vogel 2006: 456).

Non-profit biobanks open up important channels for coordinated community influence over health system governance. This type of biobanking can be 'a means for groups traditionally excluded from biomedical R&D to inspire research on their ailments, and to do so on their terms' (Malinowski 2005: 58). Moreover, disease advocacy groups are precisely the type of community organization that the draft report specifies as crucial to both deliberations on and implementation of a large-scale national cohort study. A national cohort, however, would be poised between actively seeking strong input from interest groups while also representing the general welfare. Classical pluralism assumes that freedom of association, and competition among interest groups, will produce optimal policy outcomes in a free society. However, the political history of the health sector in the twentieth century reflects the tendency of organized interests in modern societies to split between elite and grassroots members, and the power of money to distort competition in favor of the most well-funded and/or well-connected groups. In an historical analysis of failed reform movements for universal healthcare, B. Hoffman argues that 'the tradition of pluralism or incrementalism in American health politics [has] generally been seen as an impediment to large-scale reform' (Hoffman 2003: 79). Classical pluralism, of course, *intends* to make sweeping political change rare, with limited government valued much more highly than efficient government. In the contemporary healthcare sector, this means that highly visible and well-organized groups such as the AIDS activists of the 1980s and 1990s can be effective in changing health policy with respect to specific research priorities (such as anti-viral drug research), particularly when a health issue or social movement captures the sustained attention of the US mass media. Large-scale health reform, potentially including a multi-generational project such as the proposed national longitudinal cohort, is much more difficult to achieve in a fragmented political system.

The NIH acknowledges that 'at its most encompassing, the public is everyone outside the NIH' (NIH 2004b: 9). In addition, 'the availability of increasing amounts of health information makes everyone rare in some way' (Kohane and Altman 2005: 2074). For a national cohort to be successful in the United States, proponents must craft effective regulations out of the likely clash between the prospective study of common diseases, pressure from narrowly-focused disease advocacy organizations, the rights of the individual patient, and the general public interest in the integrity of the health system's priority-setting processes. This could be exceedingly difficult in a contemporary political climate where 'what began as de Tocqueville's "web of voluntary associations"' evolved by the late twentieth century into a cacophony of interest groups, each 'with very narrow foci representing an elite with the power and financial resources to create effective lobbying organizations' (Cooper *et al.* 2006: 77).

Conclusion

In a review essay on research in health administration, J. White concludes 'the fundamental public administration question [is] legitimacy – on what grounds may some officials be allowed to exercise power over citizens?' (White 2007: 174). National biobanks must be constructed in both a material and figurative sense, and individuals – the source of the genetic and clinical information upon which the edifice of genomics depends – must be enrolled either through coercion or persuasion to donate tissue samples and share their medical and genealogical histories. The regulatory controversies arising from the storage and use of these data differ substantially across nation-states, depending on such factors political culture, institutions, and the relative power accorded to stakeholders from the public and private sectors. What individuals and societies perceive to be important is not stable either within or across states, while the search for regulatory legitimacy is both ongoing and bounded by institutions. As I argued previously with respect to the Estonian Genome Project:

> Most democratic states proclaim similar policy goals with respect to biotechnology, such as participating in the knowledge economy, improving public health, and promoting innovation. However, states differ substantially on the policies they adopt, the amount of risk they accept, and how they manage the relationship between public research and private sector commercialization.
>
> (Fletcher 2004: 4)

In other words, they differ in terms of governance. A large-scale national cohort study could invigorate public trust in the American healthcare system. Indeed, governance *of* a national biobank, what Gottweis and Petersen refer to in the introduction to this volume as 'the goal to integrate biobanks into the pre-existing fabric of regulation, law and society', opens a policy window within which governance *through* the deliberations on a national cohort could direct the market–state–society relationship towards a more participatory and universal model of healthcare. However, the enduring American tensions between private and public incentives in the healthcare sector, and between interests groups and the general public welfare, mean that the 'politics of legitimation' (Salter and Jones 2005: 711) in the American biobank case is, fundamentally, the politics of building and maintaining public trust.

References

Association of American Medical Colleges (2006) 'Comments on the SACGHS draft report', available online at www.aamc.org/advocacy/library/research/corres/2006/080206.pdf (accessed 3 November 2006).

Cambon-Thomsen, A. (2004) 'The social and ethical issues of post-genomic human biobanks', *Nature*, 5: 866–73.

Collins, F. S., Green, E. D., Guttmacher, A. E. and Guyer, M. S. (2003) 'A vision for the future of genomics research: a blueprint for the genomic era', *Nature*, 422: 835–47.

Couzin, J. (2005) 'Gene bank proposal draws support – and a competitor', *Science*, 309: 684–5.

Cooper, T. L., Bryer, T. A. and Meek, J. W. (2006) 'Citizen-centered collaborative public management', *Public Administration Review*, 66: 76–88.

DeMarchi, B. and Ravetz, J. (1999) 'Risk management and governance: a post-normal science approach', *Futures*, 31: 743.

Faster Cures (2006) 'Saving lives by saving time'available online at www.fastercures.org/home.php (accessed 1 June 2006).

Finger, M. (2005) 'Conceptualizing e-governance', *European Review of Political Technologies*, 1: 1–7.

Fischer, J. (2001) '21st century gold rush', *ASEE Prism Online: Exploring the Future of Engineering Education*, available online at www.prism-magazine.org/jan01/gold_rush/gold_rush.cfm (accessed 28 May 2006).

Fletcher, A. L. (2004) 'Field of genes: the politics of science and identity in the Estonian Genome Project', *New Genetics and Society*, 23: 3–14.

Genes and Public Policy Center (2006) 'NHGRI funds center's assessment of public attitudes about population-based studies of genes and environment', news release 29 September. Online. Available www.dnapolicy.org/news.release.php?action=detail&pressrelease_id=62 (accessed 3 November 2006).

Genetic Alliance Biobank (2004) 'Genetic Alliance Biobank launched' (news release 27 October), available online at www.biobank.org/launch.asp (accessed 5 February 2007).

Hall, S. (2003) 'Personalized medicine's bitter pill', *Technology Review*, 106: 62–70.

Hoffman, B. (2003) 'Health care reform and social movements in the United States', *American Journal of Public Health*, 93 (1): 75–85.

Kohane, I. S. and Altman, R. B. (2005) 'Health-information altruists: a potentially critical resource', *The New England Journal of Medicine*, 353 (19): 2074–7.

Lyall, C. (2005) 'Workshop summary: new governance tools for new technologies', paper for the Innogen Workshop, ESCR Genomics Policy and Research Forum, University of Edinburgh. Available online at www.innogen.ac.uk/Events/Workshops/New-Governance-Tools-for-New-Technologies-Innogen-Workshop (accessed 1 June 2006).

Malinowski, M. J. (2005) 'Credits, debits and population health futures', *The Journal of Law, Medicine and Ethics*, 33: 54–69.

Marshfield Clinic (2006) 'Personalized Medicine Research Project', available online at www.marshfieldclinic.org (accessed 1 June 2006).

National Cancer Institute (2006a) 'TCGA: how will it work?', available online at http://cancergenome.nih.gov/media/process_textonly.asp (accessed 7 February 2007).

National Cancer Institute (2006b) 'Consortium of cohorts', available online at http://epi.grants.cancer.gov/Consortia/cohort.html (accessed 31 July 2006).

National Children's Study (2006) 'e-Update October 2006' available online at http://nationalchildrensstudy.gov/news/e-updates/e_update_102006.cfm (accessed 3 November 2006).

National Human Genome Research Institute (2005) 'NHGRI policy roundtable: the future of genomic medicine', available online at www.genome.gov/17516574 (accessed 1 June 2006).

National Institutes of Health (2004a) 'Request for information: design and implementation of a large-scale prospective cohort study of genetic and environmental influences on common diseases', available online at http://grants.nih.gov/grants/guide/notice-files/NOT-OD-04–046.html (accessed 3 November 2006).

National Institutes of Health (2004b) 'Enhancing Public Input and Transparency in the National Institutes of Health Research Priority Setting Process', available online at http://copr.nih.gov/reports/enhancing.pdf (accessed 5 February 2007).

National Institutes of Health (2005) 'NIH launches comprehensive effort to explore cancer genomics', available online at http://cancergenome.nih.gov/media/news.asp (accessed 1 June 2006).

The New Atlantis (2003) 'Bank on it: Britain constructs a universal genetic database', *The New Atlantis*, Fall: 102–3.

Nye, J. S., Zelikow, P. D. and King, D. C. (eds) (1997) *Why People Don't Trust Government*. Cambridge, MA: Harvard University Press.

Personalized Medicine Coalition (2004) 'Personalized Medicine Coalition fact sheet', available online at www.personalizedmedicinecoalition.org/communications/pmc_factsheet.pdf (accessed 7 February 2007).

Price Waterhouse Coopers (2005) *Personalized Medicine: The Emerging Pharmacogenomics Revolution*, available online at www.pwcglobal.com/extweb/pwc publications.nsf/docid/3B11C597C386E52385256FB1006F945B (accessed 1 June 2006).

Putnam, R. (2000) *Bowling Alone: The Collapse and Revival of American Community*. New York: Simon & Schuster.

Rosen, C. (2003) 'Liberty, privacy and DNA databases', *The New Atlantis: A Journal of Technology & Society*, 1: 37–52.

Salter, B. and Jones, M. (2005) 'Biobanks and bioethics: the politics of legitimation', *Journal of European Public Policy*, 12: 710–32.

Secretary's Advisory Committee on Genetics, Health, and Society (2006) 'Policy issues associated with undertaking a large US population cohort project on genes, environment, and disease', draft report, available online at www.od.nih.gov/oba/SACGHS/reports/LPS%20Public%20Comment%20Draft%20Report.pdf (accessed 3 November 2006).

Skloot, R. (2006) 'Taking the least of you', *The New York Times*, available online at www.nytimes.com/2006/04/16/magazine/ (accessed 28 May 2006).

Speth, G. and Haas, P. (2006) *Global Environmental Governance*. Washington, DC: Island Press.

Tait, J. and Mittra, J. (2004) *Complexity and Innovation in the Pharmaceutical Industry*. Innogen Policy Brief, December 2004.

Verband Forschender Arznelmittehersteller e.V. (2006) 'The pharmaceutical market', available online at www.vfa.de/en/ (accessed 31 May 2006).

Vogel, G. (2006) 'Biobank ties cancer genes to rare developmental disorder', *Science*, 311: 456.

White, J. (2007) 'Politics and (health) administration', *Public Administration Review*, 67 (1): 174–8.

8 Governance by stealth

Large-scale pharmacogenomics and biobanking in Japan

Robert Triendl and Herbert Gottweis

By any definition, *Biobank Japan* is among the world's largest and best-funded efforts to build a large-scale biorepository linking biological materials, DNA samples, and data on genetic variation and clinical information taken from actual patients. At the same time, Biobank Japan is one of the least debated or controversial of the large biobank projects currently being developed. This chapter will discuss this apparent paradox in Japanese biobank governance and map the transformative potential of this 'invisible project' within the context of Japanese biomedicine.

Launched officially in June 2003, Biobank Japan has received more than US $200 million over a three-year period from the Japanese Ministry of Education, Culture, Sports, Science, and Technology (MEXT). In less than three years, more than 200,000 blood samples and medical records from over 170,000 patients have been obtained and are presently stored in anonymized form at the Japan Biobank facilities located at the Institute of Medical Sciences at the University of Tokyo (IMSUT). Among similar publicly funded efforts to build large-scale repositories linking clinical and genetic information obtained from diseased individuals, Biobank Japan stands out in terms of its size, overall budget, investment in facilities, and perhaps also the sheer speed with which the project has been implemented, patients enrolled, and samples collected. It constitutes a complex network of governance in which a large number of different actors cooperate, exchange information, and interact on a regular basis.

Within a relatively short time, Biobank Japan has become one of the largest standardized collections worldwide – and certainly the largest collection within Japan – of blood samples linked to patient disease histories obtained with the explicit purpose to store samples, extract DNA, medical records, and genetic information and provide them to users in medical research. In addition, although few samples have been genetically analyzed, the project's leader, Yusuke Nakamura, has a solid reputation in the area of rapid genotyping; with this background, it is likely that Biobank Japan will be at the forefront of research in the actual genotyping and eventual analysis of the thousand of samples obtained for each of the forty-six diseases covered by the project.

Most remarkably, despite its size and the unprecedented amount of funding provided by the Japanese education ministry, Biobank Japan remains largely unknown to the Japanese public and, by some measure, even to the country's medical and biological research community. It seems that somehow Biobank Japan has bypassed many of the political-regulatory issues that have surfaced so strongly in comparable projects elsewhere. At the same time, Biobank Japan constitutes a forceful intervention in biomedical research in Japan and aims to reconfigure medical practice towards a model of 'personalized medicine'. This intervention is warranted, proponents explain, by the specificities of the 'Japanese genome', and the argument links to a discourse of national identity and biological citizenship.[1] Although Biobank Japan was being set up rapidly, its precise goals and strategies were far from clear and are still in a process of being specified. Biobank Japan is, not unlike the other biobank projects discussed in this book, a 'machine to make a future', designed to give answers to questions that its designers have not yet fully developed (Rabinow and Dan-Cohen 2005: 4). At the same time, Biobank Japan is an infrastructure that transforms and intervenes in the ways life is being governed in Japan. Biobank Japan, thus, is a good example for how deep transformations in the governance of life do not necessarily lead to broader social contestation. In this chapter we will show how biobank governance 'by stealth' operated in Japan.

Biobank Japan and its sociopolitical context

Biobank Japan has been launched in a rapid fashion and without any serious consultation or a formal process with the wider scientific or ethical community, a fact that contrasts strikingly with the long and protracted process that preceded the launch of Biobank UK (see Masui *et al.* 2002; Corrigan and Petersen, Chapter 9). Media attention to the project, although increasing, has been fairly limited, and many of the first press reports on the project were written by journalists who are known as partisan supporters of the project's leader, Professor Yusuke Nakamura. And although a small army of self-declared academic analysts ponder over every small detail of the Biobank UK effort, Japanese social scientists or ethicists concerned with the Biobank Japan are rare; neither have any famous anthropologists from outside of Japan declared Biobank Japan as their new 'field' yet. One is certainly tempted to explain this apparent lack of public interest in the Biobank Japan project in terms of certain particularities of Japanese society and culture or what some, in Japan and abroad, see as the still somewhat underdeveloped 'civil society' in the country (Schwartz and Pharr 2003).

But there is nothing intrinsic in Japanese culture or society that can account for the lack of public attention to Biobank Japan, and, in fact, the smooth operation of Biobank Japan is far from self-evident. For example, Japan has a well-developed system of voluntary blood donations, but many leaders in the scientific community believed before the launch of Biobank

Japan that collecting a large number of samples from the Japanese people would be all but impossible. In addition, the Japanese public is interested in what scientists actually *do* with the human materials collected from Japanese patients.

In fact, over the past decade informed consent procedures have been a hotly debated topic, and although much remains to be done, significant changes have occurred over the past few years in the way patients are enrolled in experimental research programs (Leflar 1996; Feldman 2000). Neither is Japan a society free of stigmatization based on heredity. For example, atomic-bomb victims and their heirs (the *hibaku-sha*) have long been subject to various forms of social discrimination and stigmatization that, in more subtle forms, continue to this day (Morioka Todeschini 1995, 1999). The same is true for families afflicted by genetic diseases (Muto *et al.* 2000), or even entire groups of the Japanese society (such as the *buraku-min*, the Japanese outcasts). In each of these cases (and in many others that could be added), certain individuals or societal groups are discriminated against in various ways based simply on heredity. In summary, based on the preceding evidence, to assume that the Biobank Japan project stirred little debate simply because issues of informed consent or participation in genetic research are not deemed problematic or because no receptive public exists is hardly convincing. Whereas the planners and funders of the Biobank UK project have voted for a lengthy consultation process, mostly confined to a specialized audience of experts in genetics, epidemiology, or bioethics, and lasting several years, in the case of Biobank Japan, the opposite seems to have been the case.

As we will show, the sociopolitical shaping of Biobank Japan was characterized by a specific politics of identity that operated in a double move: First, the various scientific-technological activities involved in the development of Biobank Japan were systematically presented and interpreted as being part of the building of the genomics research and infrastructure in Japan, positioning Biobank Japan as a national technology and competitiveness project. Second, this particular aspect of genomics research was to be devoted to the creation of a system of personalized medicine in Japan. Because of the unique features of the 'Japanese genome', strategic investments and corresponding activities by the state were essential in building pharmacogenomics in Japan. This bio-nationalist definition of 'Japanese identity', of what was at stake in collecting and analyzing blood samples in Japan, the construction of the project as essentially linked to the 'nature of the Japanese genome' and as being an exercise in (basically non-controversial) 'personalized medicine' were key factors resulting in the 'public invisibility' of the biobank project, and worked against the raising of ethical and political issues that were so prominently addressed in other biobank development projects.

Building the Biobank Japan network

A unique aspect of Biobank Japan is its integration into the Japanese genomics research infrastructure, its 'clinical definition', and its positioning as a project of overcoming biotechnology market failure and promoting the international competitiveness of Japan. The Biobank Japan project is, in many ways, built upon earlier investments into research on genetic variation, Single Nucleotide Polymorphisms (SNPs), and pharmacogenomics. It is this gradual shaping of a system of structured cooperation that is at the core of building a network for the governance of Biobank Japan. While some of these earlier efforts, funded at equally generous levels as Biobank Japan, were of fundamental importance to innovations, the Biobank Japan project aims to 'realize' pharmacogenomics and to bring it to the clinic. Generous state support was crucial, so the narrative went, because the state had to make up for 'market failure', the negligence of Japanese industry in relation to pharmacogenomics. As we will see, this was the central narrative explaining the relevance, urgency, and rationale of Biobank Japan, and thus a key tool in its governance.

This politics of SNPs operated in a specific context. The promise of an entirely new class of powerful 'personalized' therapeutics was of considerable political attraction in a country with a rapidly aging population. In addition, Japan consumes large amounts of pharmaceuticals and suffers what appears to be a relatively high number of drug-induced fatalities; new approaches such as pharmacogenomics did promise to considerably reduce drug side effects. In summary, research on SNPs and pharmacogenomics fit well into the official discourse of promoting or, as some government officials would put it, 'reorganizing' the pharmaceutical industry. Perhaps most important, as one politician put it when addressing an audience of actual and would-be biotechnology entrepreneurs at a public forum at the University of Tokyo in 1999: 'This is an area where Japan can still make it.'

The strong emphasis on SNPs research in Japan was also reinforced by Japan's research funding system and its definition of central challenges for technology policy making. During the early 1990s, Japan's pharmaceutical industry remained highly fragmented, with a large number of mid-sized, family-run companies and a biotechnology industry dominated by large companies focused on process technology, rather than recombinant DNA technologies or genomics. The continuing stagnation of the Japanese economy triggered a number of efforts to reinvigorate technical change and innovation through government spending on public sector research as well as more focused interventions targeting specific industries. During the second half of the 1990s, life sciences and biotechnology emerged, again, as a major funding focus for public-sector research spending. In 1999, the five ministries and agencies involved in biotechnology and life science funding released a joint declaration and statement of intent on biotechnology R&D policy (Triendl 1999a). This document, largely an effort by METI officials, urged for more coordination in research policy and more research funding and prepared the ground for an increased focus on biotechnology and life sciences research in overall government funding. This was given particular impetus by the government's Millennium Project – a special appropriation within the year 2000 budget for science and technology that provided significant funds for research on SNPs and pharmacogenomics (Triendl 1999b). In fact, in the budget proposals for 2000, released in August 1999, SNPs research already figured prominently. The sudden rise of SNPs research as a main priority in life sciences funding at practically *all* ministries and agencies (with the exception of the agricultural ministry) is remarkable. Certainly, the launch of the SNPs Consortium by the US National Institutes of Health together and a group of pharmaceutical companies, announced at about the same time, was an important influence, and Japanese government agencies and media certainly took note of this event.

One result of the joint agencies' declaration was increased spending for research on genetic variation in Japan that funded several dozen research groups at significant levels and led to the creation of several new research centers. Two leading geneticists, the evolutionary geneticist and leader of the Japanese DNA databank Takashi Gojobori and the surgeon turned geneticist Yusuke Nakamura, were elected as the main scientific leaders for the biotechnology efforts under the Millennium Project. Total funding made available for research on genetic variation during 1999–2004 was likely in the range of US $350–450 million, and an important fraction of this funding went to efforts run by Nakamura to scan the entire human genome for genetic variations. In a major effort that included significant investments in infrastructure, Nakamura's group at the University of Tokyo identified well over 200,000 SNPs and built what was then the world's fastest genotyping platform, located at the RIKEN SNPs Research Center.

Already during the 1990s, both METI and the Japanese health ministry had launched a number of research consortia in the field of genetics, including a major effort to build large-scale cDNA collections in Japan, rather than to sequence the whole genome (Japan's contribution to the international genome projects was fairly modest). Because of this effort, some Japanese government officials have portrayed the move to study small genetic alterations and polymorphisms as simply the next logical step. Perhaps more important, about that time SNPs emerged as a priority for industry research and cooperation between the public sector and industry became a significant goal.[2] SNPs mapping was expensive and large scale, and for a country with a research funding system notoriously understaffed, the lure of 'large scale' must have been an important attraction. Here was an activity that needed extensive funding, in the order of hundreds of millions of US dollars, that was clearly oriented towards advancing the pharmaceutical industry into the age of genomics.

The 2003 science budget in Japan included a number of high-profile funding efforts at the various ministries to fund research in priority areas termed 'Leading Projects' at the Ministry of Education, Culture, Sports, Science, and Technology (MEXT) and 'Focus 21' projects at the Ministry of Economics and International Trade (METI). Both schemes reflect subtle changes in science and budget politics, notably the creation of the Council for Science and Technology Policy (CSTP).

Biobank Japan was constructed in this changing, and as yet somewhat unstable, environment for research funding as a follow-up project to the large investments in SNPs research at the University of Tokyo. Hence, Biobank Japan was not created as an isolated project but built into the larger research infrastructure in genomics research and framed as a crucial part of a larger overall biotechnology strategy in the interest of the nation.

The story of Biobank Japan

An explanation of Biobank Japan's development must focus on those dimensions of governance that were critical to creating the conditions for the operation of the project. Instrumental was a particular narrative mobilized in representing the project to the various Japanese publics. While the eventual goal of the Biobank Japan effort is to periodically collect samples from some 300,000 individuals suffering from forty-six common diseases and to provide blood samples, DNA samples, medical records, and genetic information to scientists working on relevant diseases in an open process, Biobank Japan was not represented as an infrastructure project or as an effort to build a large 'biobank'. Rather, the project was defined as an aspect of a large pharmacogenomics project funded by the Japanese education ministry. While the term 'Biobank Japan' is now occasionally used even in government documents, this term was never used as the project's official name. Instead, the effort was initially presented as the 'made-to-order' medicine realization project' and is now officially termed the 'project to realize medical care that applies genetic information from individual patients'. This framing had, as we will see, significant implications for biobank governance.

The project was officially called 'Personalized Medicine Realization Project', a name that revealed something about the way education ministry officials presented the project to the finance ministry (which must ratify any funding decisions in Japan) and to politicians (who serve as lobbyist, supporting project proposals by agencies or ministries to the finance ministry). Although launched at a time when privacy and data protection issues emerged as major topics on the political agenda, policy makers also avoided a protracted ethical debate by presenting Biobank Japan as a pharmacogenomics project aimed at the development of better medicines with fewer side effects. Interestingly, few newspaper articles covering the project use the term 'Biobank' at all, and most describe the Biobank Japan project strictly as an undertaking in personalized medicine. The only location where the term 'Biobank' is used in a public website of the project, geared to patients rather than researchers, is in the site's URL. In practically none of the public relations materials used to promote the project is the term 'Biobank' used; neither do media reports employ the term frequently.

As mentioned earlier, the idea of personalized medicine had a special appeal in Japan, as it was linked to the idea of the specificity of the Japanese genome. A survey of newspaper articles using keywords such as 'personalized medicine', 'made-to-order medicine', 'biobank', or 'Iressa' confirms this fact: the term 'Biobank' is rarely mentioned and when it is, its references are usually to the UK project. Interestingly, the public debate surrounding the lung-cancer drug Iressa, which caused over 200 cases of drug-induced death in Japan over 2002–3, not only helped to promote the concept of pharmacogenomics but also demonstrated that significant differences exist in drug responses between Japanese and Asian populations and Caucasians, and that these differences are likely caused by certain mutations that are more common in Asian populations.[3] Thus, the Iressa case pointed to the importance of the pharmacogenomics concept – especially in the case of cancer therapy – and further demonstrated the necessity of countries like Japan to run their *own* pharmacogenomics studies using Japanese patients, rather than to simply rely on results produced in the West.

The business of biobanking: 'correcting market failure'

Economic arguments have been crucial in building political support for personalized medicine and pharmacogenomics projects. An important fraction of activities in the United States or even Iceland were undertaken by companies – such as Celera Diagnostics, Perlegen, or DeCode – and funded by private capital raised on the stock market or through the R&D budgets of large multinational pharmaceutical companies. In Japan in the 1990s, the mechanisms to fund start-up companies with payoff times for investors of several years or even decades were hardly in place. Japanese pharmaceutical companies lacked the instruments to build large in-house SNPs efforts by themselves and were not prepared to outsource such activities. Public funding for pharmacogenomics and large-scale sample collections in Japan was portrayed as largely a response to the country's pharmaceutical industry's failure to invest in emerging technologies such as pharmacogenomics. This narrative was shared by leading bureaucrats and politicians backing these programs. A certain amount of techno-nationalist posturing comes with such statements, and there are numerous interviews by Nakamura and others in which the tenor is simply that Japan 'must not allow the US to gain a lead in exploiting the human genome'.[4] Where the market failed, the government had to step in to protect the nation's interest.

At the same time, however, few efforts existed to build upon organizational structures and policies that would facilitate the rapid transfer of new technologies and new knowledge to industry. This constitutes a serious shortcoming in the evolving Japanese biobank governance structure. Given the large amounts of funding involved, this might seem surprising. Yet it is important to recall that the linkages between industry and academia in Japan even during the 1990s remained relatively weak, resembling in many ways the situation in many continental European countries where only during the 1990s did governments start to build novel programs to better connect universities with the corporate sector. Even large organizations such as RIKEN are only starting to reinforce partnerships with industry. While Yusuke Nakamura had been involved in a number of collaborations with the Japanese pharmaceutical industry, such as the PharmaSNPs Consortium, his relationships with Japanese companies have at times been strained. For example, in 2002 pharmaceutical companies accused Nakamura of 'selling off' research that had been publicly funded in Japan to US start-up companies after the University of Tokyo licensed some of its findings, whereas Nakamura countered that he had attempted to negotiate a deal with Japanese companies but had been rebuffed.

Building the biobank

A key role in launching Biobank Japan was and is played by Yusuke Nakamura. Working first at a research institute funded largely by the city of Tokyo, Nakamura was quickly offered a position with the Institute of Medical Sciences at the University of Tokyo (IMSUT), a leading research center in the medical sciences where since the mid-1990s another returnee scientist, the immunologist Ken-ichi Arai, was trying to revolutionize Japanese research organizations and build an open, elite center for biomedical research that would link basic biology with clinical practice. Nakamura's status in Japan is perhaps best compared to J. Craig Venter in the United States, and Nakamura often comments that if only the financial means and circumstance were available he would have left the public sector long ago in search of a better environment to realize his ambitions.[5]

As a follow-up project to a number of large SNPs sequencing efforts at Yusuke Nakamura's lab, Biobank Japan was an effort to 'scale-up' existing activities and increase the number of samples available to Nakamura's group. In some sense, Nakamura's laboratory was the early testing ground for Biobank Japan. To study genetic variation, Nakamura's group had used mostly samples obtained from volunteers within the laboratory, a practice that is not uncommon, if not uncontroversial, in genome research – J. Craig Venter also sequenced his own genome. Yet in Nakamura's lab, which recruits large numbers of MDs from all over Japan into temporary postdoctoral positions, self-experimentation also took a more radical dimension. Many of the 100 or so staff in this laboratory also provided blood samples for the project. Biobank Japan has built its sample collection strategy through a series of personal contacts between Yusuke Nakamura and senior figures in the Japanese medical establishment.

None of the ministry's advisory bodies had been consulted on the subject of the project, and virtually no debate took place within the scientific or biomedical research community on a large-scale biobank project. The 'Tailor-made Medicine Realization Project' was eventually selected by the finance ministry from around forty 'Leading Projects' at the Ministry of Education and included in the ministry's budget proposal to the national Diet, the Japanese parliament. Still, total funding for the project had been reduced to around US $200 million, and US $70 million was to be provided through supplementary budget funding, which requires funding to be spent within a very short timeframe.[6] Some members within the Japanese scientific community, including leading geneticists, expressed surprise or even anger when the decision by the Ministry of Education to fund Biobank Japan was presented to the public in early 2003. Industry representatives and the Japanese Biotechnology Association at least privately expressed anger about the fact that industry had not been involved in the planning stages and that no convincing strategy had been presented on how to involve industry in this effort (Tsuboi *et al.* 2002). Interestingly, despite much irritation in the scientific community and the medical establishment and, to say the least, rather skeptical views from industry, there was little public debate about Biobank Japan. This is somewhat surprising, given the fact that – at that time – the Japanese parliament was discussing legislation on data privacy that foresaw a sweeping exemption for biomedical research but not medical practice.

An important move in the construction of the Biobank Japan governance network was the decision not to seek contributions from prominent public universities and medical schools, which are known for their parochialism. Biobank governance is about creating a large network of cooperation and exchange, in which the collection of blood samples for DNA analysis is a key dimension. This collection is set up rather differently from country to country. In this respect, Biobank Japan pursued a particular strategy. Yusuke Nakamura voted to enroll a number of private hospital groups and universities to participate in the project.[7] Certainly, engaging medical schools at large national universities, such as the University of Tokyo or Osaka University, would have required a complex process of consultation and negotiation that eventually would have almost certainly led to a fragmentation of project leadership. Also, building on some of the existing biobank activities in Japan would have required going through a complex process of negotiations and ethical consultations that undoubtedly would have altered the timetable of the project dramatically. Without question, in a public climate in which debates on data protection were heating up because of new legislation that was eventually passed in 2004, a public debate about a large-scale biobank project would have meant risking stalemate and opened the door to political opponents and critics of Nakamura (of which there are many). The decision to work with private hospital groups and organizations with very limited in-house activity in genomics or biobanking meant bypassing all these issues – a move that even Nakamura's critics say was crucial for the project's 'success'. Keeping the sample and information collection strategy as simple as possible appeared to have been another key component for success. For example, Biobank Japan collects only peripheral blood, rather than tissues, which can be accomplished in a relatively short period of time. Further, in some cases clinical information is entered into a highly standardized format by Biobank Japan's own staff, rather than by participating hospitals, thus eliminating many bottlenecks and problems.

Working with private hospital groups had another advantage. When asked for their motivation to engage in the Biobank Japan project, some of the medical doctors who function as local project coordinators simply stated they were 'assigned to the task by their superiors'.[8] Superiority, career perspectives, and a rather particular form of 'peer pressure' may also explain the relatively high number of participating medical doctors in most (or all) of the hospitals interviewed.

Relying on private hospitals, and in particular on one large private hospital group, may have in fact helped to increase data uniformity, but it caused numerous problems in the clinical data collection process – arguably the most crucial aspect of any Biobank effort, since it is the quality and uniformity of the clinical information available that determines what can be done with the collected data. For example, not all participating hospitals were found able to provide full data sets for all patients. Also, although some participating hospitals directly input data into the simplified data masks developed by Biobank Japan, in other cases this work is done directly by Biobank Japan project staff. Again, Biobank Japan has voted for a 'large numbers' approach and a sufficiently large sample size for all diseases covered by the project. The sample collection process in itself is, by and large, independent from the use of the Biobank Japan resources. At the level of the individual physician this is a necessity, given the sheer size of the project. Yet at the institutional level such an approach seems at least questionable. However, it is a fact that few of the organizations involved in Biobank Japan have in-house pharmacogenomics capabilities or intend to build significant R&D activities through participation in the project. In practice, it appears to have been somewhat easier to enroll organizations that did not have large in-house activities in pharmacogenomics (except in cases in which these activities were already linked to the Nakamura group). The absence of large national universities and hospitals as participants in the sample collection process is most striking, and it remains to be seen what this means for Biobank Japan as an open-use infrastructure.

Participating hospital groups have followed different approaches with respect to hiring and training coordinators: whereas some hospitals have hired dedicated staff, many rotate existing staff into a few hours of work as medical coordinator every day. All coordinators were trained directly by Nakamura's group. In the larger hospitals, where hundreds of samples are obtained every day, there is a constant flow of people into the coordinator's office, most of them aged fifty to sixty. The whole process often takes only fifteen to twenty minutes and includes a discussion with the coordinator and a five-minute video that explains the project. The rate of patients who actually make a donation is astonishingly high and, at most hospitals, rises to over 90 per cent. Some patients return home to discuss the issue with their families before making a donation, which occasionally leads to further inquiries about the project. Concerns about genetic privacy are occasionally mentioned but appear to be relatively rare. Rather, the biggest problem for most hospitals in meeting their targets is that the number of new donors slows down after an initial surge, simply because the overall number of patients is limited. Detailed records on the number of patients from each department or responsible physician are kept, and while some doctors refer literally thousands of patients, others stop after only a few. According to one physician, the few minutes needed to explain the project and hand over a leaflet 'add up'. The entire process of sample collection is also a process of aggregating numbers, and Biobank Japan keeps track of the number of patients enrolled and donations received on a constant basis. Yet few if any questions about representation and selection are asked: Rather, the goal is simply to obtain as many patients for the forty-six disease classes covered by the projects as possible. The obvious assumption is that the Japanese population is sufficiently 'homogeneous' to warrant a blind sample approach. No specific provision *at the sample collection* stage has been made to include (or exclude) minorities, such as Ainu or inhabitants from Okinawa, or else atomic-bomb victims.

Regulating and owning Biobank Japan

The joint ministries initiatives in 1999 and the Millennium Project of 2000 provided a new motivation for the various government expert committees drafting regulation for biomedical research to develop a comprehensive framework for genetic research. There have been various efforts to develop unified guidelines for genetics research, most particularly the 'Fundamental Principles of Research on the Human Genome' released in 2000 and the 'Ethics Guidelines for Human Genome/Gene Analysis Research' released in 2001. Still, Japan's regulatory approach to human materials and genetic databases remains fragmented, with several types of guidelines often covering only research funded by a given ministry or agency. Japanese commentators have also pointed out that most of the legal issues related to large biobank facilities (including informed consent procedures, sample collection and provision, reuse of existing samples, confidentiality and sample anonymization, data protection, ownership and intellectual property rights) are regulated by guidelines only, without a binding legal framework. There remain a number of issues with respect to disclosure practices in medical practice (such as the right to access one's own medical record) that are far from resolved. Thus, much is actually left to the ability of the scientific and medical community to govern itself. But earlier experiences with regulating advanced medical practices – and notably the protracted debate on organ transplants – seem to question the ability of the Japanese biomedical and medical communities to self-regulate. Furthermore, regulation *in practice* depends also to a considerable extend on the Institutional Review Boards (IRBs) at the organizations that carry out the research. In other words, the government leaves all practical issues with respect to the implementation of these guidelines to the organizations that actually *perform* the research, with government agencies intervening only in cases of significant problems.

Hence, in the case of Biobank Japan, the planning and implementation of the project was almost entirely in the hands of Yusuke Nakamura's group with very limited interference from MEXT. Despite the size of the project, no planning group within MEXT was set up; neither did the government attempt to form a special advisory group for the project, nor in any other way impose an external governance regime on the project. Instead, Biobank Japan is in many ways run like research projects that receive a tiny fraction of the funding that Biobank Japan receives. In such a scenario, it is perhaps only logical that MEXT also left it to Yusuke Nakamura and the University of Tokyo to work out any ethical or legal issues related to this undertaking. Because of the way Biobank Japan emerged, as an aspect of building a genomics research infrastructure and never narratively positioned as a 'biobank project', a bioethical assessment of the project was at best an afterthought. Public debate on the project never materialized, and the emerging regulatory measures followed the logic of other, much smaller life science research projects.

Since 2003, a bioethics committee headed by a legal scholar involved in many government bioethics committees became part of the governance structure of Biobank Japan. Earlier, the same committee had been defined as an 'ELSI project' attached to Biobank Japan but was then given the more formal status of a bioethics advisory and review board. This change came after protests by the JMA to a study in which a scientist had contracted the work to collect lifestyle data for a cancer cohort study from ordinary citizens undertaken in Kumano-cho near Hiroshima with no obligation to protect confidentiality (Normile 2003). Accordingly, the agenda of the bioethics committee was from the start limited to the data collection process.[9] The group has made visits to all sites where samples are collected and has provided the Biobank Project with a continuous assessment of the sample collection process. In a sense, Biobank Japan has pragmatically used the Bioethics Committee as an additional means to supervise the sample collection process at participating hospitals, and the publicly available reports of the Bioethics Advisory Board clearly indicate this fact. In many ways, the reports read almost like summaries of a site review or an inspection. One report indicates that 'there is not enough space in the consultation room to guarantee privacy when several consultations are on-going', while in another questions are raised about whether the video equipment is installed properly and whether 'prospective donors can really hear the explanations on the video well'. In most cases, the issues raised by the committee ('Can a consultation room located in a glass-cubical at the entrance of the hospital really guarantee privacy of donors?') are somewhat superficial and rather removed from some of the key questions facing a large biobank project. While arguing that the Bioethics Committee has followed a somewhat 'bureaucratic agenda' and is of only limited use to the project, Nakamura also points out that individual members of the committee have helped considerably to finetune the sample collection process and the relationship with hospital partners and clinical coordinators (Nakamura 2003). Still, in its official role, the Bioethics Committee is perhaps little more than yet another layer in a well-calibrated public relations approach – its role is not to provide an overall assessment of the project but to help Biobank Japan better delineate the project from the participating hospitals and to help to contain the risk of inadequate approaches to sample collection at the participating hospitals.

In addition to various technical factors and project design choices, organization factors are also imposing a number of important limitations on the Biobank Japan project. Surprisingly little attention has been paid to issues of management, use, and ownership of the resources collected. When the Biobank Japan project was formally announced, some of the most basic questions about ownership were yet to be worked out – and this situation persists to this date (Nakamura 2003). At present, the facilities and resources of Biobank Japan belong to and are owned by the Japanese government but are managed by the University of Tokyo. The situation with respect to intellectual property rights is similar but somewhat more murky, and the absence of clear rules reflects the fact that companies do not have access to Biobank Japan at this stage. Innovations emerging from the project are patented by the respective organizations performing the research. Although the project is now approaching the end of its first term, the future of ownership still remains unclear. One possibility (and one that appears to be favored by Nakamura) is to transfer the management of the Biobank Japan resources to an existing bioresource facility, such as the RIKEN Bioresource Center (BRC). It is not apparent whether such a move would help make the resources of the project more broadly available, since the distribution infrastructure at BRC remains weak. An alternative would be to transfer Biobank Japan to a nonprofit organization managed by the University of Tokyo. Yet maintaining a sophisticated biobank facility has its cost, and such a move without assurance of continued long-term funding might put the collection in jeopardy. Companies have not been a main target as users for Biobank Japan. Still, uncertain ownership issues have made it only more difficult for companies to actively engage in the project. Companies in Japan have tended to be skeptical about the project, and one industry representative interviewed has argued that the 'lack of transparency' at the beginning of the project made it extremely difficult for industry to participate. Other industry representatives have pointed to the fact that Nakamura is himself involved in a biotechnology start-up and there is a sense that any output of the entire effort would 'first of all benefit Dr Nakamura's own company', an accusation that seems largely unfounded. Still, unsettled questions about project governance, ownership, and long-term strategy have clearly impacted the relationship between the Biobank Japan project and Japanese industry – a fact that appears in striking conflict to the way Biobank Japan has presented itself.

Conclusion: Biobank Japan, governing the future

Biobank Japan constitutes an aggressive state-led push to fund industrial biotechnology in Japan. Yet it was not conceived and presented to the public as an infrastructure for either future medicine (as was the case in Iceland or Estonia) or biomedical research (as was the case with Biobank UK). Rather, Biobank Japan was initially conceived and presented to the public as a necessary tool to realize the vision of made-to-order 'medicine', the industrialized version of personalized medicine, a key project in the context of the nature of the 'Japanese Genome'.

Because of the way Biobank Japan was set up, it largely failed to enroll collaborators in national universities and university hospitals, including groups that had been involved in large-scale cohort studies for many years or even for many decades. Rather, in order to be operable and to collect samples, the project had to recruit a number of private hospital groups (at least some of which were using the project as a means to promote a new image for themselves as innovators in medical research). And because of the opposition from the medical establishment and the research community, Biobank Japan also had to find novel ways to connect with patients and patient groups. Yet, despite its size, Biobank Japan was never defined to be a Japanese national biobank (even if, in the end, it may come close to such an effort) and, even today, is not positioned in this fashion.

The Biobank Japan project is radical in its approach – politically as much as scientifically. What distinguishes it from many other biobank efforts is the strong emphasis on personalized medicine as the explicit agenda of the project. In the dominant policy narrative, Biobank Japan is about the development of innovative approaches toward personalized medicine, rather than about providing biological resources to scientists. And personalized medicine has been represented as a highly specific challenge for Japan, as a strategy to 'defend' the 'Japanese genome' and provide adequate healthcare for modern Japan.

However, the history of the Biobank Japan project also defines its future governance challenge: Having successfully built up a major biorepository in a very short time, the main issue for the future of Biobank Japan is how to put this resource to use and to multiply linkages with the user community in both academic research and private industry. The construction of Biobank Japan has implied the avoidance of many of the contentious issues and questions typically raised about similar projects. It was guided by the idea to keep the project 'invisible'. There was nothing 'natural' or 'automatic' about the success of this strategy. In fact, even in Japan, one could easily imagine a project like Biobank Japan becoming trapped in a complex debate about the project's intrinsic scientific value, genetic privacy, or ownership of the resources collected. But while enabling the successful launch of a large biobank project in a very short timeframe, this approach seems to have had its cost. Clearly, the way the project has been conceived and launched affected some of the most basic assumptions and strategic choices for collecting and analyzing samples and data. While enabling the rapid launch of the project in an initial phase, some of these choices may limit the scientific impact of the project in the long-term. Perhaps most important, its close association with an applied pharmacogenomics project that has followed a radical scientific design is limiting the number of research groups in Japan that can make use of the resources collected. Whereas the reliance on private hospital groups (rather than large university hospitals at national universities) has made the project possible, it has done very little to integrate other scientists in the effort. Finally, limited interaction with the private sector in the planning and execution phase of the effort means that, despite its stated objective of correcting 'market failure' through government intervention, Biobank Japan will have only a very limited effects on the Japanese pharmaceutical industry. Thus, Biobank governance by stealth, it seems, might come at significant costs concerning its economic and scientific impact.

Notes

1 'Made-to-order Medicine Realization Project', or 'ôda-mêdo iryô jitsugen purojekuto' in Japanese, was the name under which the project was first launched. The name reflects the fact that an earlier large-scale effort was made around order-made medicine. For sheer convenience, we will use the name 'Biobank Japan' throughout.

2 The SNP Consortium (TSC) launched by Arthur L. Holden in cooperation with the US National Institutes of Health and the pharmaceutical industry is still seen by many as a rare example of a successful large-scale collaboration between the public sector and private industry (Hill *et al.* 2004).

3 The story of Iressa (Gefitinib) in Japan is particularly intriguing in relation to the Biobank Japan project. The announcement by AstraZeneca in July 2002 that its lung-cancer drug had been approved in Japan, thus making it, in effect, the first of a new class of cancer drugs to reach the market, was widely reported and boosted the company's stocks. However, the drug was not taken from the market. Iressa was approved a year later in the US after it was found that adverse effects in the US are significantly lower than in Japan (Dyer 2003). Rokuro Hama, a well-known critic of safety regulation in pharmaceuticals in Japan, has argued that the effective rate of adverse effects was probably even higher (Hama and Sakaguchi 2003). Several groups in Japan and the US, including Yusuke Nakamura's group, have subsequently identified possible reasons for the difference in response to therapy with Iressa by US and Japanese patients, which also suggest that the drug may be effective in only 10 per cent of the patient population (Paez 2004).

4 See, e.g., an interview with Nakamura in *Nikkei Biobusiness* (2002), 17: 69. Published in September 2002. It would appear from the release of the budget plans by the various ministries that the function of such interviews is often simply to provide support for budget proposals.

5 Nakamura is also the founder and external director of the company OncoTherapy (www.oncotherapy.co.jp, Kawasaki, Kanagawa Prefecture), with over 3.5 billion Japanese yen in paid-in capital and 54 staff – arguably one of the best-funded Japanese biotech start-up companies.

6 Supplementary budget funding is highly political and is typically used for buildings, facilities, or instrumentation.

7 The hospitals participating in Biobank Japan are Iwate Medical University, Osaka Medical Center for Cancer and Cardiovascular Diseases, Cancer Research Foundation, The Tokushukai Group, Tokyo Geriatric Medical Center, Juntendo University, Japan Medical University, and Nihon University.

8 The following quotations were obtained from interviews undertaken with supervising medical doctors and, in one case, a supervising medical coordinator at participating hospitals. All interviewees were guaranteed anonymity. Transcripts of the interviews are on file with the authors.

9 Within Biobank Japan, all samples are freshly collected and, at present, Biobank Japan stores only samples that have been collected within the project; there have been no efforts whatsoever to include existing collections in the Biobank Japan framework. While individual scientists connected to the Biobank Japan may still use such collections, it was a deliberate choice not to include existing sample collections to keep the project 'clean' and avoid any potentially controversial issues. While the 'Basic Principles' do not allow the reuse of human samples for purposes not covered in the initial informed consent process, the 'Ethics Guidelines' of 2001 differentiated among three types of samples and provide for the reuse of certain categories only.

References

Asian Technology Information Program (2003) *The Biobank Japan Project*, ATIP03.042. Report available online at www.atip.org (accessed 6 October 2006; document on file with authors).

Branscomb, L., Kodama, F. and R. Florida (eds) (1999) *Industrializing Knowledge: University-Industry Linkages in Japan and the United States*. Cambridge, MA: MIT Press.

Coleman, S. (1998) *Japanese Science: From the Inside*. London: Routledge.

Dyer, O. (2003) 'FDA announces fast track approval of new drug for lung cancer', *BMJ*, 326: 1004.

Feldman, E. A. (2000) *The Ritual of Rights in Japan: Law, Society, and Health Policy*. Cambridge: Cambridge University Press.

Hama R. and Sakaguchi, K. (2003) 'The Gefitinib story', *ISDB Newsletter*, 17: 1. Available at www.npojip.org/english/The-gefitinib-story.pdf (accessed 9 October 2006).

Hill, D. E., Brasch, M. E., del Campo, A. A., Doucette-Stamm, L., Garrels, J. I. *et al.* (2004) 'Academia-industry collaboration: an integral element for building "omic" resources', *Genome Research*, 14: 2010–14.

Hiroi, Y. (1992) *Amerika no iryôseisaku to nihon: kagaku, bunka, keizai no interface* (*US Health Policy and Japan: At the Interface Between Science, Culture, and the Economy*). Tokyo: Chikuma Shôbô.

Hito soshiki no ishoku-to e no riyô no arikata ni kan-suru senmon iinkai (2000) *Hito soshiki no ishoku-to e no riyô no arikata ni tsuiteio Hito* (*Guidelines for the transplantation of human tissues*). Tokyo: Ministry of Health and Welfare.

Kneller, R. W. (2003) 'University–industry cooperation and technology transfer in Japan compared with the US: another reason for Japan's economic malaise?', *University of Pennsylvania Journal of International Economic Law*, 24: 329–449.

Kôsei kagaku shingikai sentan iryô gijutsu hyoka bukai (2000a) *Ekigaku-teki shuhô o mochiiru kenkyû-tô ni okeru kojin jôhô no hogo-to no arikata ni kan-suru chôsa kenkyû* (*Considerations on Data and Privacy Protection in Epidemiological Research*). Tokyo: Ministry of Health and Welfare.

Kôsei kagaku shingikai sentan iryô gijutsu hyoka bukai (2000b) *Idenshi kaiseki kenkyû ni fuzui suru rinri mondai-to ni taio suru tame no shishin* (*Guidelines Concerning Ethical Issues Related to Human Genome Research*). Tokyo: Ministry of Health and Welfare.

Kôsei kagaku shingikai sentan iryô gijutsu hyoka bukai (2000c) *Soshiki bank jigyô o tsujite hito soshiki no ishoku-to e no riyo no arikata ni tsuite* (*Rules for the Transplantation of Human Tissues Using Tissue Banks*). Tokyo: Ministry of Health and Welfare.

Kôsei kagaku shingikai sentan iryo gijutsu hyôka bukai, hito-soshiki o mochiiru kenkyû kaihatsu no arikata ni kan-suru senmon iinkai (1998) *Shujutsu-to de teishutsu sareta hito-soshiki no kenkyu kaihatsu no arikata ni tsuite* (*Rules for the research use of human surplus tissues from surgical interventions*). Tokyo: Ministry of Health and Welfare, Health Policy Bureau.

Kôsei-shô (2000). *Idenshi kaiseki ni yoru shippei taisaku zoyaku ni kan-suru kenkyu ni okeru seimei rinri mondai ni kan-suru chosa kenkyu* (*Research on Bioethics Issues in Human Genome Research*). Tokyo: Ministry of Health and Welfare.

Leflar, R. (1996) 'Informed consent and patients' rights in Japan', *Houston Law Review*, 33: 1–112.

Macer, D. (1992) 'The "far east" of biological ethics', *Nature*, 359: 770.

Mandavilli, A. (2004) 'Profile Yusuke Nakamura', *Nature Medicine*, 10: 560.

Maruyama, E. (2000) 'Hito genomu kenkyû, idenshi kaiseki ni kan-suru saikin no seifu shishin' ('On Recent Guidelines for Genetics and Genomics Research'), *Jurisuto*, 1193: 49–55.

Masui, T. (2002) 'Hito no koto ha hito deiu jidai no naka de: jintai-yûrai shiryô ni izon shita iryô seibutsu-gaku kenkyû no kiban-seibi nit suite' [original English language title: 'On the required infrastructure of the current use of human materials and information in biomedical research in Japan'], *Rinshô Hyôka*, 30.

Masui, T., Sofuni, T., Ishü, M., Imanishi, Y., Hideaki *et al.* (2000) 'Kôsei saibô banku ni okeru hito soshiki, saibô toriatsukai rinri mondai e no torikumi' ('Approach toward ethical issues regarding human cell and tissue use at the JCRB Cell Bank'), *Tissue Culture Research Communications*, 19: 1–15.

Morioka Todeschini, M. (1995) *Hiroshima 50 ans.* Paris: Autrement.

Morioka Todeschini, M. (1999) 'Illegitimate sufferers: A-bomb victims, medical science, and the government', *Daedalus*, 128: 67–100.

Muto, K. *et al.* (2000) 'Nihon no idenbyô kenkyuu to kanja, kazoku no kea ni kan-suru chôsa' ['Research on genetic diseases and patient and family care in Japan: a survey'], *Studies*, 4.

Nakamura, Y. (2003) Interview with authors, Institute for Medical Sciences, University of Tokyo, 6 October.

Normile, D. (2003) 'Japan guidelines under fire after protest halts study', *Science*, 301: 1039.

Nudeshima, J. (2001) *Sentan iryô no rûru: jintai riyô wa doko made yurusarerunoka* (*Rules for Advanced Medicine: To what Extent Should the Use of the Human Body Be allowed?*). Tokyo: Kôdan-sha.

Paez, J. G. (2004) 'EGFR mutations in lung cancer: correlation with clinical response to Gefitinib therapy', *Science*, 304: 1497–1500.

Pechter, K. (2001) 'Measuring the University-Industry Linkage in Japan: System Assessment for Innovation', unpublished doctoral dissertation, University of Tokyo.

Porter, G. (2004) 'The Regulation of Human Genetic Databases in Japan', *SCRIPT-ed*, 1: 491. Available online at www.law.ed.ac.uk/ahrb/script-ed/issue3/japan.asp (accessed 6 October 2006; document on file with authors).

Rabinow, P. and Dan-Cohen, T. (2005) *A Machine to Make a Future: Biotech Chronicles.* Princeton, NJ: Princeton University Press.

Riken Advisory Council (2004) *Strengthening Scientific Governance in an Age of Change, Report by the 5th RAC*. Wako: RIKEN, The Institute for Physical and Chemical Research. See also www.riken.jp/engn/r-world/info/report/rac/index.html (accessed 12 December 2006; report on file with authors).

Salter, A. J. and Martin, B. R. (2001) 'The economic benefits of publicly funded basic research: a critical review', *Research Policy*, 30: 509–32.

Schwartz, F. J. and Pharr, S. J. (eds) (2003) *The State of Civil Society in Japan*. Cambridge: Cambridge University Press.

Triendl, R. (1999a) 'Japan reorganizes national biomedical research groups', *Nature Medicine*, 5: 365.

Triendl, R. (1999b) 'Millennium money for Japanese scientists', *Nature Medicine*, 5: 1095.

Tsuboi, E., Kawahara, N., Mitsuishi, T., Oshima, A. and Yonemoto, S. (2002) 'Data security is crucial for Japanese science', *Nature*, 417: 689.

Upham, F. K. (1996) 'Privatized regulation: Japanese regulatory style in comparative and international perspective', *Fordham International Law Journal*, 20: 396–511.

Part 3

Biobanks, publics and citizenship

9 UK Biobank

Bioethics as a technology of governance

Oonagh Corrigan and Alan Petersen

Introduction

In March 2007 the first participants were recruited to UK Biobank and the world's largest planned national repository of human DNA and health-related data to be established for epidemiological research was officially launched. The establishment of an Ethics and Governance Framework (EGF) and an independent ethics council to oversee the project along with attempts to elicit the views of a range of stakeholders to inform this process prompted an independent international review panel to claim that the 'UK Biobank's approach to ethical governance was exemplary and would be held up as a gold standard across the world' (MRC 2007).[1] To be sure, the long awaited launch of the project had been preceded by a lengthy period of planning and ethical deliberation. When proposals for a UK genetics population database first emerged in 1998 it was apparent to those involved in discussion about the setting up of such a project that careful attention would need to be paid to its ethical aspects. In particular, following on from the disputes and controversies that had thwarted the Icelandic genetic database (see Pálsson, Chapter 3) the UK's funding bodies were aware of the need to pay special attention to the requirements of informed consent and to ensure the public acceptability of the project. In the Icelandic case controversy had arisen over both its departure from the normal practice in research of gaining specific informed consent from participants and in its coalition with the industrial sector. In the UK case the issue of consent was to be carefully deliberated upon, and it was decided that although the resource could be accessed by commercial entities this would be subject to adherence to strict ethical protocols. Furthermore, the biobank project would be a public venture funded by UK medical charities and government departments[2].

The UK also had a long and successful history in managing large scale prospective health population studies where there had been little, if any, previous contention on the grounds of their ethical acceptability. Nevertheless, a number of events had occurred during the 1990s in the UK science and medical domains that had challenged science policy and developments in genetics and medicine. In particular, the UK's public disquiet about authorities' response to the mad cow disease (BSE) and the GM controversy, combined with a series of medical scandals, in the late 1990s and early 2000s (e.g. Alder Hey, Bristol Royal Infirmary, the Harold Shipman affair) are widely perceived as evidence of a failure of regulation in the fields of biomedicine and biotechnology that has engendered mistrust. Evidence of mistrust is seen to be revealed by surveys, such as that undertaken by the Office of Science and Technology and the Wellcome Trust in 2000 which revealed respondents' 'amazement' in science, but their 'low level of confidence in regulation and the Government' (Office of Science and Technology and The Wellcome Trust 2000). Concerns about a 'crisis of trust' have been echoed in various reports such as the Science and Society report of the House of Lords Select Committee on Science and Technology (2000). Consequently, in recent years, scientists and policymakers have been keen to develop means of exploiting the opportunities offered by new technologies without undermining public trust. The engendering of trust – or rather the management of mistrust – has become a central issue for contemporary risk governance (Barnett *et al.* 2006). Therefore despite previous success in facilitating and managing large-scale prospective population studies, many of which had already begun to collect DNA, the funders were aware that the proposed UK Biobank project would need to gain the support not only of the half a million proposed participants but also the population at large. In this light, the framework deployed for governing previous projects was seen as an insufficient governance mechanism to ensure public confidence and thus ultimately the success of the project. The need to find new measures that went beyond traditional frameworks was echoed in international ethics debates where they were given a sense of urgency (Greely 2001; Joly and Knoppers 2006; Knoppers and Chadwick 2006). In particular, processes such as the reliance on the implementation of informed consent procedures and analyses of the risks and benefits to prospective research participants were seen as ill-fitting and inadequate for this new breed of national biobanks.

The concern over genetic databases must also be understood in relation to the wider attention given to the ethics and governance of genetics more generally. Bioethics in particular has acquired an increasingly prominent role in relation to the 'new genetics'. The Human Genome Project and the 'genetic revolution' has since the 1990s been accompanied by debates and deliberations on the ethical and social implications of such developments. In the UK, as elsewhere, the genetics revolution has brought with it an ethics revolution. While in the US 5 per cent of public funds spent on genetics was used to fund ELSI work examining the ethical, legal and social implications of developments in genetics, funding institutions in the UK established various programmes of research into the ethical and policy implications of genetics and specially designated commissions were established. For example, during the 1990s the Wellcome Trust established a biomedical ethics programme of research specifically to examine issues relating to developments in genetics and the neurosciences, and the Nuffield Council of Bioethics[3] was established to identify, examine and report on the ethical questions raised by recent advances in biological and medical research, predominantly those relating to genetics.

It would be easy to get the impression that bioethics has only been given birth to since this period but this is not the case. Bioethics as a mode of governance has been increasing in significance since the latter half of the twentieth century (Rothman 1991; Bosk 1999; Stevens 2000). Any examination of issues relating to the governance of biobanks must begin by examining the role played by bioethics more generally as a mode of governance. In this chapter we argue that bioethics needs to be understood as a contemporary mode of governmentality, entailing a range of techniques aimed at identifying risks, rendering them calculable and manageable. Following Michel Foucault (1991) and his followers (e.g. Nicholas Rose (1999) and Mitchell Dean (1999)), we use governmentality to refer to the rationalities of governance or 'the conduct of conduct': the ways of thinking and acting involved in prevailing systems of rule. Increasingly, these involve administrative techniques of risk management; the deployment of preventive interventions oriented to populations deemed to be 'at risk' or presenting 'a risk' in relation to, for instance, illness, crime, unemployment, anxiety, and so on (Castel 1991). These may include surveillance technologies (e.g. monitoring devices, CCTVs, biometric cards), the routine diagnostic testing of those deemed to be 'at risk' (e.g. of giving birth to disabled children) or contracting particular diseases (e.g. 'late-onset' conditions, workplace related diseases) and, increasingly, the 'profiling' of whole 'at risk' or 'risky' populations. As this chapter will demonstrate, the ascendance of genetic explanations of health and illness, particularly as manifest at the level of populations (as in public health genetics and preventive medicine), marks a shift in emphasis for bioethics and the mechanisms of governance. Whereas technologies of governance were previously aimed at identifying and managing the risks related to the health and welfare of individual participants in discrete, time-limited research projects, currently, the object of risk calculation is now more 'the public' insofar as 'it' has the capacity to undermine or jeopardize a project. 'Society', or at least societal response to UK Biobank (as with other endeavours in genetics) has become one of the major preoccupations of new modes of bioethics governance. This 'social turn' (Corrigan in press) in bioethics is not without its tensions and contradictions and the previous mode of bioethics has not disappeared but rather this new mode has been charged with the additional task of engendering societal support for research. Increasingly, 'ethics' involves anticipatory strategies oriented to engendering public consent and legitimacy for projects whose goals and modus operandi have in the main already been established. The challenge of engendering consent is particularly acute in the case of public health and preventive medicine given that research may involve very large populations involving the collection of data over a long period of time (potentially decades). As we conclude, the governmental implications of this shift in emphasis in bioethics needs to be recognized, especially the potential for adopted strategies to engender mistrust and resistance.

A genealogy of bioethics as a mode of governance

As already mentioned, biobanks are not unique in their attention to ethical concerns and in the deployment of bioethics as a mode of governance. Contemporary developments in medicine and science are accompanied by increasing discussions, debates, guidelines and activities associated with bioethics. The role played by bioethics, as argued by Raymond De Vries (De Vries 2004), is to define the moral terrain of medicine and science, identify 'ethical problems' and provide solutions for them. However, bioethics is a heterogeneous field taking on multiple forms and from a Foucauldian perspective bioethics must be understood in its broadest sense as a discourse – a way of knowing, a form of knowledge and practice as evidenced in everyday uses of the term, its deployment in policy guidelines, in the establishment of oversight committees and commissions, and its representation in academic debates and media discussions. It is manifest in bioethics guidelines such as professional codes laying down principles and practices for the conduct of biomedical research as well as being enshrined in legislation such as laws that ban reproductive cloning. Other manifestations include various oversight and research ethics committees, special designated commissions, such as the aforementioned Nuffield council on Bioethics and the UK's Human Genetics Commission. Bioethics also presents itself as a form of expert knowledge located within the academy with increasing numbers of bioethics scholars and professional bioethicists being trained and practising in hospitals and as advisers to policy makers (Elliott 2002). As an academic field, it has largely been dominated by analytic moral philosophy and law with the social sciences playing a minor role (Fox and Swazey 1984), however as concern has arisen over the need to gain societal approval of research, social scientists have recently come to play an increasingly significant role (De Vries, Turner *et al.* 2006). For example, the Wellcome Trust' biomedical ethics programme was specifically aimed at funding multidisciplinary empirical-based ethical and social research. The main task of research ethics committees has become one to ensure that the informed consent process is facilitated (Holley and Foster 1998) and the assessment of risk has become primarily the task for individuals taking part, with emphasis being placed on sufficient written information about the relevant risk and benefits and subsequent understanding by participants in forming decisions regarding participation. Indeed, informed consent can be understood as a tool for the 'individualisation of risk' (Beck 1992; Bauman 1998). From a Foucauldian governmentality perspective then, bioethics as a mode of governance can be understood as one in which practices are aimed at emerging 'risks' in the field of biomedicine.

According to conventional research bioethics guidelines, risks in large scale prospective research of the kind associated with biobanks have been viewed as 'minor' compared to those categorized as 'major', such as ones involving risk of serious physical harm. This is because large scale prospective cohort studies are largely observational with interventions limited to surveys and the collection of blood or maternal tissue samples, with little direct harm anticipated. When compared to the risks involved in the participation of a clinical trial, for example, where risks could include physical harms such as drug-induced side effects, participation in population-based non-interventionist studies have been viewed as less risky. Although more recently the ongoing nature of such projects has been viewed as problematic insofar as meaningful consent is harder to achieve and risks cannot always be determined at the outset, these studies were not, at least initially, deemed particularly problematic. The MRC's National Survey of Health and Development British 1946 (MRC 1946) birth cohort study, for example, was facilitated and has continued since with little reflection on its ethical consequences. The study, which has involved the follow up of over 16,000 babies born in a single week during 1946, has since mapped biological and social pathways to health and disease. Now aged sixty-one years the original population has been studied at least twenty-one times, with follow up studies involving a vast array of surveys and home visits, and more recently the collection of DNA. Indeed data has already been collected on genes concerned with respiratory health (www.nshd.mrc.ac.uk/genetics.html). The lack of concern accorded these and other kinds of biobanking activities in the past was not something limited to the UK context. As Hirtzlin *et al.* have revealed in their survey on biobanking in Europe, explicit consent for biobanking activities as such is a recent issue and written consent was not always obtained in the past. Their study reveals that informed consent procedures for medium size and smaller scale repositories of tissue used for research had in the past often not been carried out and where informed consent forms were used they were generally only based on information relating to the primary use envisaged and the long-term uses were not always mentioned (Hirtzlin *et al.* 2003). Although informed consent has been a fairly well-established practice since the 1970s, this has been based more on the randomized control trial model of research rather than prospective population studies.

Given the ongoing nature of prospective studies and their openness in terms of the range of social and health related issues under study, there has been a growing awareness for the need to more carefully assess their particular ethical issues. Risks of harm to participants in the form of potential breeches in privacy and infringement of rights have also begun to be acknowledged. Also questions have arisen about the extent to which individuals could be truly informed when the implications for those involved in such research cannot be known in advance. By the 1990s concern about consent in relation to cohort studies began to be expressed. In particular, problems with the fact that it is parents (usually the mothers) of the eventual subjects that give 'surrogate' consent on behalf of their as yet unborn children were raised. In recognition of this a specially designated ethics committee was established in the 1990s to advise on the ethical issues associated with the Avon Longitudinal Study of Parents and Children, a study involving 15,000 pregnant mothers following the subsequent birth and development of children born to them.[4] In addition, the Wellcome Trust funded a project designed to improve ethical understanding, including the problems posed in gaining and maintaining consent of the children involved (Williamson, Goodenough *et al.* 2004) In particular, the study aimed to understand the ethical issues in epidemiological research, with special reference to clinical genetics, to longitudinal studies, and to research involving children.

As mentioned, by the 1990s public concern had arisen over data privacy and policy formations such as the new Data Protection Act (1998) were introduced in the UK to safeguard personal information collected and stored pertaining to individuals. This was coupled with growing public concern and policy responses to the storage and disposal of human tissue and body parts where 'public trust in the medical profession and the conduct of medical research had been seriously eroded' (Martin and Kaye 2000). Following media and public concern over the removal, retention and disposal of human organs and tissues after post mortem examination at the Royal Liverpool Children's Hospital, Alder Hey in 1999 an inquiry panel was appointed to investigate these issues. This led to the formation of the new UK Human Tissue Act (2006). These changes demanded new consideration be given to the nature of risks in the collection, storage and retention of human tissue, especially when accompanied by the storage of personal and health related data. Furthermore, and perhaps more significantly it was clear that it was important to regain public support and trust, and a new emphasis was placed on ensuring public acceptability for such projects. Risks then in the form of public mistrust have become new objects for bioethics governance. This was echoed in other areas of scientific development where there was a particular concern among scientists and those lobbying for developments in genetics in the UK that public opinion should not undermine endeavours in human genetics, as had been the case for GM foods. Indeed, as was made clear in a statement by Alan Doyle, scientific programme manager for the Wellcome Trust, 'the [UK Biobank] project must be acceptable to the public' for it to be a 'good investment' for the UK Biobank company (Commission 2002). The role for ethics then was to ensure the support of the public because rejection by the public would result in a 'bad' scientific and financial investment. The ethical gaze then shifted from one concerned primarily with balancing the risks to participants and the benefits to 'society' to one where 'society' itself became the focus of ethical intervention.

To be sure, traditional ethics concerns relating to the risks posed to participants are acknowledged but these are addressed as non-physical risks relating to 'privacy, confidentiality and reputation'. As mentioned earlier, such risks are conventionally viewed within research ethics guidelines as minor in comparison with risks entailed as a result of interventions that affect physical health and wellbeing. Risks to participants in UK Biobank were discussed during one of the Ethics and Governance Council meetings in 2005. Risks such as those associated with taking a blood sample of an individual, being asked a question during the assessment centre visit that they are uncomfortable answering . . . and of a breach in confidentiality' (UK Biobank Ethics and Governance Council 2005) were identified as 'minimal'. The necessity for introducing a robust ethics and governance framework in relation to UK Biobank has arisen then not in relation to the likely seriousness of risks to potential research participants but rather in recognition of the importance of establishing public support in order to keep such a project viable. Assessment of risks marks a shift away from the primary concern with making calculable and managing the risk of harm to *participants* per se towards a concern with managing *public awareness and response.* The risks being managed in UK Biobank have less to do with the potential seriousness of harm to participants than with the threat to the success of the project itself, reliant as it is on the participation of 500,000 individuals and the wider societal support necessary for the long term continuation of the project. Ethics and governance strategies in UK Biobank need to be understood as endeavours designed to gain societal support as much as they are ones concerned with minimizing risk and facilitating informed consent. The emphases on informed consent and risk/benefit analyses have not been removed, but the story of the ethics and governance of UK Biobank reflects a shifting emphasis in what is deemed problematic for bioethics.

UK Biobank: from risk to participants to risk of 'the public'

Activities surrounding the ethics and governance of UK Biobank have taken multiple forms and, as already illustrated, take place in the context of new modes of ethics and governance surrounding the collection of human tissue and research involving genetics. The initial formal ethics and governance mechanism specifically designated to regulate the UK Biobank project was the establishment of an ethics Interim Advisory Group (IAG) in 2003. In turn, this group was responsible for setting up the Ethics and Governance Framework (EGF) in 2006. The IAG was chaired by a Consultant in Health Policy and Ethics with the remaining nine members all having some form of expertise in one of the following: bioethics, law, social science, genetics and philosophy. The IAG was also charged with giving advice to UK Biobank's funders on the best ethical practice, which was to be designed to provide a sound basis for fostering public trust and confidence in the project. The remit of the IAG and one that was later to be reflected in the Framework document was as much about minimizing the risk to the project of public rejection and ensuring public trust as it was in ensuring the risks of harm to those participants involved in the research are minimized. The

Ethics and Governance Council (EGC) is now permanently established to act as an independent guardian of the UK Biobank's Ethics and Governance Framework and to report to 'the public' about UK Biobank. Reporting to 'the public' is a key requirement placed upon the Ethics and Governance Council. Seeking public support as a basis of ethical practice is not without its critiques within mainstream academic bioethics. Bioethics guidelines are given epistemological legitimacy insofar as they have traditionally been premised on moral theories where the determination of moral acts have little to do with what people think is right but rather are determined by principles derived from theories and ethical principles. For a Kantian moral philosopher for example, moral acts are seldom ones people would choose but rather they are moral duties placed upon individuals. For utilitarianism too, moral acts are always determined by their consequences and are based on a form of moral reasoning that carefully balances the potential risks and benefits. Such balancing though is not thought to be easily done by non experts and the felicific calculus as conceived by Jeremy Bentham was envisaged as a logarithmic tool to help decisions such as those of the judiciary and was one best carried out by experts. Traditional analytic moral philosophy, upon which much of bioethics has stemmed, would not be interested in seeking the views of the public in ascertaining whether something is or is not ethical. Moral philosophers claim that you cannot determine a moral decision or a moral act based on what is and argue that to derive an 'ought' from an 'is' would be to commit a 'naturalistic fallacy' (Searle 1964). Although in recent years theories such as Rawls' concept of deliberative democracy (Rawls 1996)[5] sets forth a basis for the democratic involvement of citizens in decision making as the basis for a just and moral society, in the UK the analytic tradition of Kantian deontology and Mills' utilitarianism dominates the bioethics academic forum. A bioethics framework that is asked to formulate ethical practice while at the same time ensuring public support represents a major shift in modus operandi of bioethics in the UK. Indeed the inclusion of social scientists to participate at all in bioethics has been viewed as problematic (López 2004) in the UK and elsewhere resulting in what some have described as an ethics turf war (Hoeyer 2006).

The Ethics and Governance Council are charged with reporting publicly on the ethics and governance of UK Biobank and further changes to the emphasis placed on the objects of risk being governed is evident in the EGC's annual report 2004–5. Here the authors acknowledge that while technically UK Biobank is a prospective cohort study similar in design to those already carried out over the past fifty years, they concede that 'the context in which UK Biobank will operate invites a re-appraisal of the ethics and governance of such projects' (UK Biobank Ethics and Governance Council 2005) (UK Biobank Ethics and Governance Council 2005: 3). In particular, they cite 'public awareness' and changing perceptions of the public as well as a change in the research climate itself relating to genetics and databases (ibid). Nevertheless, there are other signs of tensions between the new focus on an ethics and governance mechanism based on public accountability and engendering societal support and the autonomy of UK Biobank in relation to carrying out its business unimpeded.

Constraints on public accountability

While in many ways the setting up of a special council seems to emphasize a genuine commitment to ensuring that the project conforms to the highest possible ethics and governance standards, there are questions about the actual powers of the Council and of the IAG that had preceded it. While the EGC is established as an independent body it has no direct powers of oversight nor has it any rights to veto or overrule any proposed actions by UK Biobank. The extent to which the EGC should have the power to veto or overrule had previously been discussed by the ethics Interim Advisory Group:

> The Group realised that powers at the latter extreme could conflict with the legal authority and responsibility of the Board of Directors, and probably would not be acceptable to the Funders or the Board. So the EGF [Ethics and Governance Framework] presents a model that depends on integrity, competence, mutual goodwill, and public visibility.
>
> (UK Biobank Interim Advisory Group on Ethics and Governance 2003)

However, some members of the IAG were at least initially sceptical of what they saw as mere rhetorical powers and thought that the EGC should have the power to exercise a veto, 'as well as independence, in order to give it authority to exert control over UK Biobank's actions if necessary and foster public perception of its protective status' (UK Biobank Interim Advisory Group on Ethics and Governance 2003). Furthermore, while UK Biobank has stated its explicit aim to garner public opinion and support, an earlier public consultation exercise commissioned by the MRC (People Science and Policy Ltd 2002) reported that most participants were eager to see an independent oversight body established, one able to set sanctions where necessary. UK Biobank's decision not to award the Council a right to veto decisions made by the company seems somewhat at odds with its desire to secure public support.

Despite the declaration that UK Biobank would be subject to transparency in a climate of openness and public accountability with reports and developments to be regularly presented and updated on UK Biobank's website, it appears that here too some compromises and tensions are evident. For example the IAG was instructed by UK Biobank to 'treat the proceedings of the IAG, the questions referred to it and any written or verbal communications between the IAG and the Funders as confidential' and not to disclose such information to third parties, including the media, without the prior agreement of the Funders (ibid). Also, at another meeting where UK Biobank's Chief Scientific Officer was present the issue concerning the UK's Freedom of Information Act (2000) was discussed. The meeting's notes report the EGC was not subject to the Act but the report on this meeting suggests that there were some concerns relating to confidentiality. Such anxieties about others being able to access 'private' documents on UK Biobank suggest that despite a rhetoric of openness and public transparency there was a sense of unease among UK Biobank members about the extent of this openness. Discussions about whether the meeting's proceedings should be held in public have also taken place. At an EGC meeting held in January 2007 (UK Biobank Ethics and Governance Council 2007) the EGC discussed whether their meeting should be open to the public.

> Strong support was expressed for the principle behind allowing observers to attend meetings (as a means of promoting transparency and trust). However, some members questioned whether allowing observers to attend its meetings is the most effective means of public engagement given other methods available.
>
> (UK Biobank Ethics and Governance Council 2007)

As we will now illustrate, public accountability extends to the creation of mechanisms of 'public engagement' that have been adopted as a kind of risk management strategy rather than as a genuine attempt to involve publics in shaping the overall aims, direction and management of the project or to open debate about the project's value and implications. A conception of 'the public' as ignorant and in need of education about the project and its aims is evident in published reports on the project's 'consultations', which reveal the lingering influence of the so-called 'public deficit' model of public understanding.

'Public engagement' in practice[6]

From the outset, the project's partners have made much of their efforts to 'consult' 'the public' and pertinent stakeholders, including in relation to the ethics and governance framework, through various means. These include panels and workshops involving members of 'the general public' from across the UK and specific groups (e.g. people with disabilities or diseases, religious and community groups), meetings with industry and focus groups with primary healthcare workers (Petersen 2005: 281–2). As previously mentioned, regular updates of web postings, which include key documents relating to the project (e.g. ethical and governance protocols) and summaries of 'consultations', reinforce this impression of transparency and openness. However, the rationale for the use of this particular mix of methods and their adequacy has not been explained or presented for discussion through this and other means and does not appear to have been the subject to questioning by the project's partners. There is no way of knowing what has *not* been made available for public scrutiny. Consultation exercises arguably limit scope for discussion and reflection: as stakeholder oriented mechanisms they typically include only those who have expressed an interest in the issues, and their potential for expanding decision making is reliant on prior decisions about their scope (Royal Commission on Environmental Pollution (RCEP) 1998). A 'public consultation' report on 'the ethical and management issues surrounding the proposed BioBank UK project', commissioned by the Wellcome Trust and MRC and undertaken by an independent public policy consultancy (People Science and Policy Ltd), provides no explanation for the rationale for the selected methods for the research (a series of group discussions with the proposed target population (45–69) and for how the sample was selected, and offers no assessment of the level of confidence that can be ascribed the findings and recommendations given the small sample size and its lack of representation from all socio-economic groups and proposed recruitment regions within the UK. The fact that the sample is not representative is, however, acknowledged in the report (People Science and Policy Ltd 2002). The text of the report indicates that the 'project team' was on hand to 'explain' participants' various 'misunderstandings' and 'confusions' (e.g. about the representativeness of the sample group) and offer 'reassurances' (e.g. regarding the monitoring of health), which suggests that the process was as much about educating the participants as an attempt to garner wide-ranging views on the project and its value. Further, many of the resulting recommendations are about ways to improve participation rates and communication processes rather than about addressing substantive concerns raised about the project (e.g. 'Taking part must be made as easy as possible. People may look for excuses to justify their inertia'; 'Setting out how this study adds value to existing work will sway some waverers. This needs to be clear in the initial recruitment material'; 'The role of GPs must be clarified and publicised') (www.ukbiobank.ac.uk/docs/consultation.pdf) (accessed 27 February 2007). In other words, the consultation reveals a conception of 'the public' as ignorant and anxious and in need of more or better or more clearly presented information; in short, a version of the 'public deficit' model in evidence in a number of other areas of science and technology communication (see Irwin and Michael 2003: 20–3). The tone of the consultation document, including the wording of questions, responses to concerns raised, the kind of recommendations made and the use of language such as the above, suggest that the process was tightly managed, allowing little scope for deviation from the predefined agenda. Such 'consultations' could have provided an opportunity for exploring and gaining feedback on substantive issues about the project: whether it should proceed at all; whether the benefits outweigh the risks; who is likely to ultimately own the data and benefit from the findings; and so on. However, it is clear from the consultation report that such questions were not explored in any detail and that the exercise was never intended to encourage debate about such issues. At the time of writing, there is no evidence to indicate that there has been a move away from this model of 'engagement' or that project's partners have reflected on the purpose and adequacy of the approach. The suitability of UK Biobank's public consultation has itself been the subject of severe criticism, including by the House of Commons Select Committee; namely:

> It is our impression that the MRC's consultation for Biobank has been a bolt-on activity to secure widespread support for the project rather than a genuine attempt to build a consensus on the project's aims and methods. In a project of such sensitivity and importance consultation must be at the heart of the process not at the periphery.
>
> (House of Commons Select Committee on Science and Technology 2003: 7)

This suggests the need for some fundamental reflection upon what the 'consultations' were meant to achieve and how methods for involving different publics in decision making and debates might be improved, assuming the project is to proceed as planned (the starting date for the recruitment of participants was delayed a number of times).

A number of potential challenges confront UK Biobank in the future as it begins to recruit and establish its database, some of which became evident during the integrated pilot phase, undertaken between February and June 2006. These include a number of practical problems such as to how to most effectively access contact details for the main recruitment phase (for a range of reasons recruitment through Primary Care Trusts proved time-consuming and difficult and it was recommended that this should be undertaken through 'just a few national points of access'); how to encourage participants to volunteer and to not to withdraw, given the periodic demands on participants' time and energies (the pilot for the project suggests that on average ninety minutes was required for the initial health assessment visit); and how to ensure an adequate response rate from people of different ages, genders, ethnicities and geographical locations (the rate of confirming an appointment to attend an assessment visit tended to be higher in older people and women) (UK Biobank Coordinating Centre 2006). In addition, a number of more fundamental and longer-term uncertainties exist, including economic conditions affecting private sector involvement (changing commercial investment priorities, with perhaps a shift away from biotechnology research – a highly feasible scenario given shareholders' demands for quick profits); healthcare workers' responses to recruitment (perhaps medical resistance, as happened in Iceland) (see Pálsson, Chapter 3); and adverse media coverage of the project or other biomedical research, which may have a 'contaminating' effect on publics' views on and support for the project. These all suggest the need for wide-ranging debate with various constituencies, including healthcare workers, commercial partners, the media and various publics. Indeed, these are issues that have been identified during EGC meetings and that are addressed in the Wellcome Trust's tender, advertised in early 2007, for research into the public's views of intellectual property and benefit sharing in UK Biobank. A particular challenge confronting UK Biobank's partners is how to make transparent to participants and publics the nature of the study and its methods, limitations and uncertainties and risks, without arousing concerns and undermining confidence in the project. The question of whether it is possible to establish genuine 'openness' while maintaining publics' confidence in the project in the longer term is debatable, since the former suggests acknowledging the many uncertainties that surround such collections, which have the potential to give rise to fears and thus resistance.

Conclusion

The emphasis on 'engagement' and 'consultation' in UK Biobank and some other contemporary biobank projects (see McNamara and Petersen, Chapter 12) seem self-evidently to be a valuable part of the process of project development. These terms have strong positive connotations, suggesting a level of public involvement in project development that few would take issue with. However, it needs to be asked what 'engagement' and 'consultation' mean in practice, in terms of allowing scope for publics to debate the substantive issues raised by such projects (e.g. the nature and the value of the science, and who ultimately owns, controls and benefits from such a collection) and to influence whether, how and under what conditions the project should proceed. Like 'transparency' the danger is that 'engagement' and 'consultation' may serve as a smokescreen for 'business as usual' or the introduction of insidious forms of control, increased bureaucratization, or practices which favour the interests of some groups over those of others (see, e.g. Cooke and Kothari 2001; Petersen and Lupton 1996). As noted, the stakeholder model of 'consultation' reflects a restricted vision of 'engagement' since it involves only those with an established stake or interest in the issues. The contemporary governmental emphasis on risk management directs attention to anticipating and developing strategies for managing adverse public responses rather than encouraging wide-ranging debate on pertinent issues. Broadening involvement in the project would necessitate consideration of appropriate mechanisms to involve publics, including those who are currently unaware of the project and its implications and those who tend to be excluded from such deliberation, and wide-ranging discussion among all constituencies about what democratic participation might look like in relation to the establishment of such a collection. It would also involve discussion on the role, limitations and implications of employing bioethics knowledge and associated practices (e.g. protocols, oversight committees, governance arrangements) in the context of biobanks. In short, the gaze needs to be turned away from 'the public' and the potential participants to the experts themselves and their knowledge and its impacts and to the conditions which give rise to mistrust. There needs to be greater appreciation of what are currently 'deficits' in experts' understanding of 'the public' (Wynne 2006).

UK Biobank's 'consultation' workshops, with their implicit 'public deficit' model of public understanding, however, indicate that thinking and practice is currently a long way from this ideal. In an effort to instrumentally engender trust, there is a strong chance that proponents of UK Biobank and other biobank projects which adopt a similar approach to 'engagement' may generate mistrust and resistance, as happened in the case of Iceland. For those interested in the governance and politics of biobanks it will be interesting to observe how various stakeholders and publics respond to UK Biobank as it begins to recruit participants and build its resource in the years ahead. As more and more people become involved as participants, researchers, and clinicians, the public visibility of the project will undoubtedly increase and consequently the level of debate. As noted, many uncertainties surround the project, including commercial involvement and the nature of research. Whether the processes and mechanisms that have been established (e.g. the Ethics and Governance Council) are perceived as adequate and the project is judged by the various publics to be of value and worthy of support and participation, remains to be seen.

Notes

1 The Joondalup Health Project in Western Australia, which provides a pilot for the proposed WA Genome Health Project, has taken a strong cue on its approaches to engagement from UK Biobank B. See McNamara and Petersen, Chapter 12.
2 The main sources of funding are from the UK's Wellcome Trust, one of the world's wealthiest medical charities, the Medical Research Council (MRC), and the Department of Health (DoH) The recruitment phase for UK Biobank is jointly funded by the MRC and the Wellcome Trust at £28m each. The DoH is providing an additional £5m and the Scottish Executive and the Northwest Regional Development Agency have each added an additional £0.5m to the total (www.ukbiobank.ac.uk/docs/Manchesterinvites.doc).
3 Since 1994, it has been funded jointly by The Nuffield Foundation, the Medical Research Council and The Wellcome Trust.
4 ALSPAC study has 'generated hundreds of millions of data points and holds half a million samples from placentas to milk teeth, the ALSPAC trove has been the subject of many collaborative studies' (Lowrence 2006).
5 The W. Maurice Centre for Applied Ethics at the University of British Columbia is currently involved in a project aimed at applying the concept of deliberative democracy to biobanking. See http://gels.ethics.ubc.ca:8213/ge3ls-arch/face-to-face/project-documents-information/documents/november-workshop/biobanks-and-deliberative-democracy-workshop (date accessed 9 May 2007).
6 Parts of this section appeared in an article previously published by one of the authors – see Petersen 2007.

References

Barnett, J., Carr, A. and Clift, R. (2006) 'Going public: risk, trust and public understanding of nanotechnologies', in G. Hunt and M. Mehta (eds), *Nanotechnology: Risk, Ethics and Law*. London: Earthscan.

Bauman, Z. (1998) *Globalization: The Human Consequences*. Cambridge: Polity Press.

Beck, U. (1992) *Risk Society: Towards a New Modernity*. London: Sage.

Bosk, C. (1999) 'Professional ethicist available: logical, secular and friendly', *Daedalus*, 128 (4): 47–68.

Castel, R. (1991) 'From dangerousness to risk', in G. Burchell, C. Gordon and P. Miller (eds), *The Foucault Effect: Studies in Governmentality*. Hemel Hempstead: Harvester Wheatsheaf.

Commission, H. G. (2002) 'Information gathering meeting on UK Biobank', retrieved 6 June 2006 from www.hgc.gov.uk/UploadDocs/Contents/Documents/Information%20gathering%20event%20on%20UK%20Biobank.doc.

Cooke, B. and Kothari, U. (2001) *Participation: The New Tyranny?* London: Zed Books.

Corrigan, O. P. (in press) 'Modes of responsibility and governance in medicine', in M. McDonald (ed.), *Languages of Accountability*. Oxford: Berg.

De Vries, R. (2004) 'How can we help? From "sociology in" to "sociology of" bioethics', *Journal of Law Medicine and Ethics*, 32 (2): 279–92.

De Vries, R., Turner, L., Orfali, K. and Bosk, C. (2006) 'The view from here: bioethics and the social sciences', *Sociology of Health and Illness*, 28 (6) (special issue).

Dean, M. (1999) *Governmentality: Power and Rule in Modern Society*. London: Sage.

Elliott, C. (2002) 'Diary: the ethics of bioethics', *London Review of Books*, 24 (23): 36–7.

Foucault, M. (1991) 'Governmentality', in G. Burchell, C. Gordon and P. Miller (eds), *The Foucault Effect: Studies in Governmentality*. Hemel Hempstead: Harvester Wheatsheaf.

Fox, R. and Swazey, J. (1984) 'Medical morality is not bioethics – medical ethics in China and the United States', *Perspectives in Biology and Medicine*, 27: 337–61.

Greely, H. T. (2001) 'Human genomics research: new challenges for research ethics', *Perspectives in Biology and Medicine*, 44 (2): 221–9.

Hirtzlin, I., Dubreuil, C., Préaubert, N., Duchier, J., Jansen, B. *et al.* (2003) 'An empirical survey on biobanking of human genetic material and data in six EU countries', *European Journal of Human Genetics*, 11 (6): 475–88.

Hoeyer, K. (2006) '"Ethics wars": reflections on the antagonism between bioethicists and social science', *Observers of Biomedicine*, 29 (2): 203–27.

Holley, S. and Foster, C. (1998) 'Ethical review of multi-centre research: a survey of local research ethics committees in the South Thames region', *Journal of the Royal College of Physicians of London*, 32: 238–41.

House of Commons Select Committee on Science and Technology (2003) *Third Report* (March), available online at www.publications.parliament.uk/pa/cm200203/cmselect/cmsctech/132/13208htm

Irwin, A. and Michael, M. (2003) *Science, Social Theory and Public Knowledge*. Buckingham: Open University Press.

Joly, Y. and Knoppers, B. M. (2006) 'Pharmacogenomic data sample collection and storage: ethical issues and policy approaches', *Pharmacogenomics*, 7 (2): 219–26.

Knoppers, B. M. and Chadwick, R. (2006) 'Human genetic research: emerging trends in ethics', *Focus* 4 (3): 416–22.

López, J. (2004) 'How sociology can save bioethics . . . maybe', *Sociology of Health and Illness*, 26 (7): 875–96.

Lowrence, W. W. (2006) *Access to Collections of Data and Materials for Health Research*. London: MRC and Wellcome Trust.

Martin, P. and Kaye, J. (2000) 'The use of large biological sample collections in genetics research: issues for public policy', *New Genetics and Society*, 19: 165–91.

The Maurice Young Centre for Applied Ethics, University of British Columbia (2007) *Biobanks and Deliberate Democracy Workshop*, available online at http://gels.ethics.ubc.ca:8213/ge3ls-arch/face-to-face/project-documents-information/documents/november-workshop/biobanks-and-deliberative-democracy-workshop (accessed 9 May 2007).

MRC (1946) 'The MRC national survey of health and development', retrieved 6 June 2006 from www.nshd.mrc.ac.uk/.

MRC (2007) 'UK Biobank: progress update', retrieved 8 June 2007 from www.mrc.ac.uk/NewsViewsAndEvents/Features/Biobank/index.htm.

Office of Science and Technology and The Wellcome Trust (2000) *Science and the Public: A Review of Science Communication and Public Attitudes to Science in Britain*. London: The Wellcome Trust.

People Science and Policy Ltd (2002) *Biobank UK: a question of trust: a consultation exploring and addressing questions of public trust.* London: MRC and The Wellcome Trust.

Petersen, A. (2005) 'Securing our genetic health: engendering trust in UK Biobank', *Sociology of Health and Illness*, 27 (2): 271–92.

Petersen, A. (2007) 'Biobanks' "engagements": engendering trust or engineering consent?', *Genomics, Society and Policy*, 3: 1.

Petersen, A. and Lupton, D. (1996) *The New Public Health: Health and Self in the Age of Risk.* London: Sage.

Rawls, J. (1996) *A Theory of Justice.* Oxford: Oxford University Press.

Rose, N. (1999) *Powers of Freedom: Reframing Political Thought*. Cambridge: Cambridge University Press.

Rothman, D. (1991) *Strangers at the Bedside*. New York: Basic Books.

Royal Commission on Environmental Pollution (RCEP) (1998) *21st Report: Setting Environmental Standards*. London: The Stationery Office. Also published in A. Irwin and M. Michael (2003) *Science, Social Theory and Public Knowledge.* Buckingham: Open University Press.

Searle, J. R. (1964) 'How to derive "is" from "ought"', *The Philosophical Review*, 73 (1): 43–58.

Stevens, M. L. (2000) *Bioethics in America: Origins and Cultural Politics*. Baltimore, MD: Johns Hopkins University Press.

UK Biobank Coordinating Centre (2006) *UK Biobank: Report of the Integrated Pilot Phase*, 19 September 2006. Stockport: UK Biobank Coordinating Centre.

UK Biobank Ethics and Governance Council (2005) *Annual Report*.

UK Biobank Ethics and Governance Council (2007) 'UK Biobank Ethics and Governance Council Tenth Meeting', retrieved 7 June 2007 from www.egcukbiobank.org.uk/assets/wtx037095.pdf.

UK Biobank Interim Advisory Group on Ethics and Governance (2003) 'UK Biobank ethics and governance framework background document', retrieved 12 February 2007.

Williamson, E., Goodenough, T., Kent, J. and Ashcroft, R. (2004) 'Children's particiaption in genetic epidemiology: consent and control', in R. Tutton and O. P. Corrigan (eds), *Genetic Databases: Socio-ethical Issues in the Collection and Use of DNA*. London and New York: Routledge.

Wynne, B. (2006) 'Public engagement as a means of restoring public trust in science – hitting the notes, but missing the music?', *Community Genetics*, 9: 211–20.

10 Biobanks and the biopolitics of inclusion and representation

Richard Tutton

Introduction

As evidenced by this volume, the question of how biobanks are established as legitimate forms of research that command support – in the form of financial investment from governmental and charitable sectors, and the willingness of individuals to volunteer their biological samples, medical records and other personal data – has become a key question. When thinking about the ways that biobanks seek to gain legitimacy, social scientists have tended to look at innovations in the arenas of governance and policy, at how a number of initiatives have held public consultations and created oversight bodies (Salter and Jones 2005), or how policymakers have used the language of 'citizenship', 'common good' or 'solidarity' to stress the value of this kind of research to society (Petersen 2005). While recognising the contribution of this work, my chapter takes a different approach, drawing on the 'co-production' framework (Jasanoff 2004) developed within Science and Technology Studies (STS). This framework suggests that we need to pay attention to biobanks as examples of knowledge-making projects as well as representing innovations in governance. This means that the legitimacy of biobanks must be understood, simultaneously, in scientific and social terms.

My argument in this chapter is that the legitimacy of UK Biobank rested on its 'inclusiveness'. In other words, the question of how and in what way ethnic minorities would be included in UK Biobank – and therefore whether they would be served by this type of research that requires significant investment of public funds – became a significant challenge that impacted on its final design. This issue of 'inclusiveness' impinges directly on matters of public trust, legitimacy and citizenship with which many social science commentators on biobanks have been concerned. I argue that the effort to ensure the 'inclusiveness' of UK Biobank required the co-production of the natural and social orders in ways that are explored here.

The inclusion and representation of ethnic minority groups in epidemiological, genetic, and biomedical research has taken on a new political significance in the last decade or so. From a history of exploitation and racism within science and medicine, so notoriously exposed in the Tuskegee studies (Reverby 2000), it is increasingly an aim of public policy to actively include minorities in research and for their potential differences to be recognised and investigated (Epstein 2007).[1] However, the inclusion of ethnic minorities in biobanks has received only limited attention. It has been acknowledged that biobanks could pose particular issues for ethnic minority groups because, by setting out to elucidate the influence of genetic and social factors on human health, these initiatives could reveal the nature and extent of differences between such groups (Chadwick 2003). Equally, how biobanks are involved in processes of ethnic categorisation, especially in relation to constructions of national identity in different countries, has also been a topic of interest (Arnason and Simpson 2003; Busby and Martin 2006; Prainsack, Chapter 13). However, the issues of ethnicity and the inclusion of different ethnic groups in biobanks remains a relatively under-researched one. For example, debates about UK Biobank have tended to focus on concerns about commercialisation (Lewis 2004), governance arrangements (Martin 2001), informed consent (Kaye 2004), and scientific rationale and design (Wallace 2005). Yet, as I show, the question of how to include ethnic minorities has been a significant part of the way the project has developed.

The chapter is divided into four main sections: I begin by outlining the co-production framework and its application to my analysis of UK Biobank. Then, in keeping with the comparative ethos of the book, I situate my particular focus on the UK in the wider international context by considering debates in the United States around the creation there of a national biobank and the inclusion of minority groups in biomedical research. This frames an outline of some key stages in the development of UK Biobank when the strategies for including ethnic minorities were discussed and formulated. This leads to the final section in which I present and analyse evidence drawn from interviews conducted with several key scientists working at UK Biobank. Through these interviews, I examine how Biobank scientists conceived of the social, scientific and practical issues of including these groups.

Biobanks and the co-production of the natural and the social

The notion of co-production has been advanced within the STS field as a way of understanding scientific knowledge and technologies that avoids both natural and social determinist accounts. Sheila Jasanoff (2004), the foremost proponent of this approach, explains that:

> Co-production is symmetrical in that it calls attention to the social dimensions of cognitive commitments and understandings, while at the same time underscoring the epistemic and material correlates of social formations. Co-production can therefore be seen as a critique of the realist ideology that persistently separates the domains of nature, facts,

> objectivity, reason and policy from those of culture, values, subjectivity, emotion and politics.
>
> (Jasanoff 2004: 3)

In particular, Jasanoff suggests that co-production illuminates how 'knowledge-making' within scientific and technological fields interacts with practices of governance in ways that mutually influence each other. This is a connexion explored by Jenny Reardon (2001; 2005) in her analysis of the controversial Human Genome Diversity Project (HGDP). She advocates that co-production is especially insightful when investigating how new techno-scientific objects come into being. Looking at how different actors describe and debate these objects one can see more easily the co-production of scientific and socio-political practices. As she notes: 'this is especially the case when the epistemological and normative implications of the emerging object are contested, and when efforts to establish its legitimacy and meaning spans multiple [. . .] contexts' (Reardon 2005: 7). Of particular relevance to this chapter, Reardon also suggests that, from the co-productionist perspective, one escapes the dichotomy of seeing concepts such as race and ethnicity as either valid scientific categories or as the products of social ideologies. Instead, as she comments, race and ethnicity do not observe the boundaries between science and society as they travel across these contexts in ways that mean they can never be used by science in a neutral way, detached from their social meanings.

Arguably, biobanks can be considered as emerging technoscientific objects, their meanings and values are contested, and they are in the process of stabilisation with some initiatives stalling and others facing an uncertain future (see Pálsson, Chapter 3). UK Biobank is a case in point: since the first proposal for what was then called the UK Biomedical Population Collection in 1999, this project has built momentum, undergone several changes in name, purpose and scope, and finally secured long-term funding. Along the way it has encountered criticism levelled at its scientific design and value-for-money and at its implications for ethical and governance practices (Kaye 2004; Petersen 2006; Wallace 2005). I also take the view in this chapter that is useful to think of ethnicity as co-produced in the sense that, as a concept, it travels across scientific and social domains in complex ways. As I show, the partners behind the UK Biobank initiative have sought to establish its legitimacy across a number of contexts by ensuring its inclusion and representation of ethnic minorities.

Biobanks, national DNA, and ethnicity

Biobanks, operating in different national and cultural settings, co-produce the natural and social orders in their own distinctive ways. It is beyond the present chapter to explore in detail the situations in nations as potentially different as Estonia, Japan, Iceland, and the Gambia all of which have sought to establish national biobanks.[2] However, a useful international comparison for the UK case is that of the United States. This is for two reasons: first that, in general, American debates about race and ethnicity have been wide-ranging and influential; second, the institutions promoting biobank initiatives in these countries talk about diversity and inclusiveness in relation to ethnic groups in similar ways.

As discussed by Fletcher (Chapter 7), the US is a latecomer to developing the idea of a national biobank. In 2004, the National Institutes of Health (NIH) put out for consultation plans to create a prospective cohort study of genes, environment and disease. This could involve up to one million volunteers. In putting forward its plans for this biobank, the NIH notes how diverse the American population is and how that must be captured by the resource in a 'representative' way. This focus on 'representativeness' reflects a concern about ensuring that findings from research using the biobank will be generalisable to the wider population, and a key issue here is that relevant social groups are included in sufficient numbers. This is why Francis Collins, the head of NIH, discussed the importance of 'over-sampling' from minorities when the biobank idea was first mooted in 2004 (Collins 2004), because sampling minorities relative to their proportion in the national population would produce too few numbers of each group to produce scientifically meaningful results. Despite this argument, the latest evidence suggests that the US project is aiming for a proportionally representative cohort of the national population, acknowledging that the project will face significant recruitment difficulties with enrolling volunteers from certain ethnic and socio-economic groups (NIH 2006).

These plans for a US biobank are being developed against the backdrop of a concerted public policy response in the United States to the under-representation or exclusion of women and minorities in biomedical research, and continuing health disparities amongst such groups (Epstein 2004; Friedman *et al.* 2000). This response prompted the 1993 NIH Revitalisation Act which mandated the NIH to include women and minority groups in clinical studies it supports as a matter of routine. As well as stressing inclusiveness, the Act also states that NIH-funded studies should be designed to permit a valid analysis of whether the variables examined affected women and/or minorities differently compared with others. Epstein (2004) argues that this legislation and the new institutional and scientific practices that it entails heralds a new 'biopolitical paradigm' in the United States, which 'traverses the boundaries between the life sciences and state policymaking' (Epstein 2004: 7), promoting an agenda of 'inclusion and difference'. It signals a move away from the 'white male model' of research studies (by which results from studies conducted on predominantly white male cohorts were extrapolated to other groups), to one centred on defining group membership and differences (see also Epstein 2007).[3]

For Epstein, this development prompts the question: on what basis should ethnic minorities be included in clinical trials or other forms of biomedical research? One reason for including people from diverse ethnic (as well as social or cultural) backgrounds is that they may have different clinical outcomes with respect to drug therapy. Only by including such people in trials in sufficient numbers, would researchers be able to investigate the extent and significance of these differences (Rochon *et al.* 1998). For example, there is a body of evidence which suggests that there are distinct ethnic variations in cardiovascular disease, in particular that people classified as 'South Asians' face an increased risk. Such variation might only be fully understood by including various ethnic groups in studies of this disease (Ranganathan and Bhopal 2006).

To meet its requirements under the legislation, the NIH adopted the racial and ethnic group classification developed by the Office of Management and Budget (OMB), which is used across the US Government. Although the OMB has explicitly acknowledged that the categories in its classification are 'social-political constructs' (Office of Management and Budget 1997), they are in effect becoming an integral part of the way that biomedical research is conducted and governed in the US. The use of such 'social-political' group categories in the biomedical context raises concerns that researchers and policymakers could end up conflating social and biological groups. This has been reinforced by the recent approval the FDA gave to BiDil® – a congestive heart failure drug marketed specifically for African-Americans – after a subgroup analysis suggested that patients who self-identified as black responded well to this drug (Carson *et al.* 1999). For some commentators in the US, this decision seems to lend weight to the idea that racial/ethnic groups are actually scientifically valid 'biological constructs' (Kahn 2004; Rathore and Krumholz 2003).

Concerns about such conflation also arise from efforts to address social and political concerns about research by developing new governance frameworks for its conduct. For example, the ethical response to the vociferous opposition to the HGDP by indigenous peoples, was to formulate notions of 'community consent' to serve as a way of consulting groups about research studies before approaching specific individuals for samples (Reardon 2005). Eric Juengst (1998) suggests that the idea of 'group rights' in this context is flawed as it creates a situation where social and biological categories are inevitably confused and conflated. He argues that the use of socially-recognised groups as 'gatekeepers' for population genetics research 'would send the wrong message, by suggesting that geneticists think that there really is a strong biological justification for the social boundaries that we draw around and between each other' (Juengst 1998: 676). This could then lead to a reification of group differences and the 'reinscribing of taxonomies of race' (Duster 2005: 1051). The NIH recognises that this danger is also inherent in its plans for a US 'biobank' (NIH 2006).

Diversity, ethnicity, and UK Biobank

The US is not alone in emphasising group differences and diversity. During the last decade or so, especially in political discourse, Britain has been celebrated as a multicultural and ethnically diverse society. When representing Britain on the international stage, for instance, the UK Government has sought to emphasise the nation's 'unique mix of cultural identities and heritages, [which] defines and adds value to contemporary Britain' (British Embassy in the United States 2006).[4] The organisation of the UK Biobank has emerged within this context, and ample evidence exists of a significant debate among those involved in the scientific and ethical/governance aspects of the project about what diversity meant – scientifically, practically and politically.[5]

Ethnicity was discussed in the initial stages of developing the UK Biobank's scientific protocol, as evidenced by a report of a meeting that took place in April 2001 which noted that:

> Ethnicity is a pivotal issue. It was suggested [during the meeting] that over-sampling should be considered to ensure sufficient numbers for analysis. However, the difficulty of properly defining ethnicity and the resources required was emphasised.
>
> (Report of the UK Population Biomedical Collection Protocol Development Workshop 2001)

When the draft scientific protocol was published, it was accepted that UK Biobank would not 'be designed to be representative of the general population of the United Kingdom' (UK Biobank 2002a). Participation would be open to everyone regardless of their stated ethnicity, so the overall cohort would likely reflect the make-up of the general population, in the sense of being – crudely – 94 per cent white and 6 per cent non-white, but representation from specific minorities would be uneven (based on 2001 census). Perhaps because of this it was decided that the larger minority ethnic groups (which, at the time, were not identified) would be targeted so that at least 3,000 subjects would be recruited from each of these groups. This aimed to permit limited studies investigating group-specific risk factors associated with differential exposures and/or genotypes.

When the draft protocol was reviewed by a panel of international scientists, they saw the 'diversity' of the British population in a positive light (UK Biobank 2002a). One noted that the 'diversity of the study population provides a definitive advantage' compared with nations whose populations were presumably more homogeneous such as Iceland (Reviewer A). Another reviewer, however, pointed out that while the diversity of the British population was a strength it could also be a weakness: if the Biobank failed to recruit sufficient numbers from all ethnic groups that would lead to an under-powered study. Interestingly, the scientific protocol was also reviewed from an ethical-legal perspective (UK Biobank 2002b). One of these reviewers asked a related question: might certain ethnic groups might be reluctant to volunteer for the study and what strategies would the scientists have for dealing with this? There was also a concern expressed about the pitfalls of ethnic classification. This reviewer cautioned against the dangers of 'pigeon-holing' subjects especially when an increasing number of people have mixed backgrounds. It was implied that a failure to recognise and accommodate this could be offensive to potential volunteers and so would undermine the project.

To focus on the issues raised by including minority groups in Biobank, a specialist Ethnicity Recruitment Sub-Group was formed. This comprised researchers directly involved in UK Biobank and drew in others with experience of recruiting and running studies with minorities. This group examined how to include ethnic minorities, in what numbers, and set out a number of recruitment measures such as translating materials into the appropriate languages and utilising 'community-based' sampling through mobile clinics and the like. However, it was recognised that sufficient subjects would not be recruited unless 'additional measures' were taken.

Therefore, using 2001 National Census data to ascertain proportionate numbers of all recognised ethnic groups in the UK, the sub-group set the overall target of recruiting 30,000 ethnic minority subjects from a total of 500,000 (approximately proportionate to the non-white and mixed population as self-reported in the 2001 Census). The 'additional measures' would focus on achieving certain numbers from select groups: Indian (14,000 – of which Hindus, 7,000; Sikhs, 4,500; and Muslims 2,500), Pakistani (6,000), Bangladeshi (2,000), and Black-Caribbean (8,000). These populations are to some extent geographically-clustered and fall within the boundaries of two Biobank recruitment centres in London and Fosse Way (which includes Bradford, Leicester and Birmingham). Therefore, these centres will be tasked with targeting these groups. We might see that the choice of these groups is partly pragmatic because of their geographical clustering. At the same time, these groups also suffer from continuing health disparities, differential rates of disease and mortality and could potentially benefit from being included in greater numbers in UK Biobank. In contrast, Chinese and Black African groups are more dispersed, so they will not be subject to 'additional measures' and will be recruited as a matter of course through all the recruitment centres.

In addition to these scientific activities, the Biobank partners also created an Ethics and Governance Framework (UK Biobank 2003). They appointed an Interim Advisory Group (IAG) – a small group of lawyers, social scientists, clinicians, ethicists and lay representatives – and commissioned it to draw up the draft version of the Framework. The IAG saw merit in recruiting a diverse range of subjects so as to 'increase the chance that the research findings eventually derived will have wide applicability' (UK Biobank 2003). At the same time, the IAG also recognised that: 'the diversity of the UK population [means that] perfect representation cannot be expected, but wide representation can' (UK Biobank 2003). Therefore, some ambiguity existed around the concept of representativeness. Following the publication of the draft Framework, the Biobank partners invited individuals and organisations to comment on its provisions. Some responses highlighted this ambiguity. One respondent in particular questioned whether the project would recruit 'UK citizens, UK residents or people who actually live in the UK' (UK Biobank 2004) a latter group that could include refugees and asylum seekers. In other words, would the project operate with specific notions of nationality or identity?

The policies adopted by UK Biobank to include ethnic minorities have drawn criticism from various quarters. Mehta – a representative of the Genetic Interest Group in the UK – and Saggar (2005) caution that in taking this approach, it 'may have limited value for minority groups, which often bear disproportionate burdens of disease' (Mehta and Saggar 2005: 207). They recommend that projects such as UK Biobank intentionally over-sample from minorities. This view is shared by the population geneticists Sharon Tate and David Goldstein (2004). Asking how genetics could help redress continuing health disparities in society, they concluded that:

> The solution is straightforward: more and better research in those groups that have been traditionally under-represented in clinical and other biomedical studies [. . .] In this spirit, Francis Collins, the director of the US National Human Genome Research Institute, recently outlined a case for [. . .] oversampling in ethnic minority groups. Similar European efforts have, however, simply ignored the issue. The UK Biobank [. . .] is sampling minorities in proportion to their representation. This effectively excludes minorities, as the numbers collected will be too small to allow identification of gene-environment interactions specific to the minority groups. We would like to see this decision reconsidered.
>
> (Tate and Goldstein 2004: S42)

They draw a comparison with the US initiative discussed above, and see that the 'over-sampling' option considered at the time by the NIH and which the Biobank did not adopt, was the best way forward. Others have been less critical of UK Biobank, arguing instead that at the very least UK Biobank should ensure it succeeds in achieving its stated goal of recruiting of 30,000 volunteers from minority populations (Ranganathan and Bhopal 2006).

The above discussion of the US and UK contexts raises a number of questions in relation to the effort to actively include all ethnic groups in research studies. On what basis should groups be included? Is there a danger of reifying differences and using socially-defined groups as proxies for genetic variation or biological differences? How should the representation of ethnic groups in studies be determined? In the rest of the chapter I aim to explore some of these questions further by drawing on a series of interviews carried out with key scientists at UK Biobank.

'The reality is [that] our genetic code is as cosmopolitan as our lifestyle': building the UK Biobank in multicultural Britain[6]

The interview data were generated as part of a collaborative interdisciplinary project between researchers in public health, bioethics and sociology of science on the current use of race/ethnicity categories in the context of 'applied population genetics' (Smart *et al.* in press). One aspect of the project has been to research how race/ethnicity is conceptualised and operationalised by various biobanks located in the UK. We successfully conducted seventeen interviews with senior research personnel at ten biobanks of varying size, design and purpose. During the interviews we explored a series of questions, including: How is race/ethnicity conceptualised? How is race/ethnicity operationalised and measured? When and how do practical scientific problems or socio-ethical concerns arise when classifying participants by race/ethnicity?

In this chapter, I draw on the five interviews undertaken with key scientific personnel at UK Biobank, which centred on the practical, scientific and social-political aspects of including ethnic minorities in this resource. My discussion of the interviews explores further how the inclusion and representation of ethnic minorities involved the co-production of natural and social orders. I begin by considering how interviewees discussed on what grounds ethnic minorities would be included, in what numbers to make the inclusion meaningful, the potential benefits that could exist for these groups from their inclusion, and which classifications were appropriate for use by UK Biobank. Through my discussion of these interviews, I show that co-production is a fruitful idiom in which to speak about how these scientists have dealt with the scientific, practical and social aspects of making UK Biobank an inclusive project.

A history of under-representation

There was a keen awareness among the interviewees of previous research practices in the fields of epidemiology, genetics and biomedicine in the UK that had produced the under-representation of ethnic minorities. One interviewee recognised that: 'Traditionally in UK studies it has been very easy to engage people of white European ancestry, particularly from like leafy suburbs or the countryside. It's been less possible to engage people from ethnic minorities' (PGD 4). This was felt to be because many researchers had not sought to do so. Thus it was felt that UK Biobank would be breaking new ground in its approach towards the inclusion of such groups. This underscored why UK Biobank needed to adopt its 'additional measures' and not rely on a random sample. As another noted:

> If we just blindly went away, closed our eyes [and] opened our eyes when we'd recruited half a million, most of the cohort would probably look a bit like me, drive the kind of car I drive, their kids go to the kind of school that I go to you know and if you like a sort of sound bite, it would be a sort of you know white, largely middle class, rural kind of population and I think that would clearly be a disservice in terms of creating a resource for the health of the UK population in terms of the future.
>
> (PGD 10)

This was supported by other interviewees who had studied the bias apparent in recruitment practices of research studies, which often led researchers to recruit subjects who were similar to themselves in socio-demographic background and not the 'hard-to-reach' groups who did not have, for example, English as their first language. This bias came about because these groups would pose more problems to researchers to recruit in terms of commitment of resources and time. Therefore, as one interviewee expressed it, 'white middle class men' (PGD 14), were the preferred option.

The grounds for inclusion

One interviewee reflected at length about the grounds on which ethnic minorities should be included in UK Biobank:

> From the information we have so far [. . .] there's nothing to say that there's going to be greater genetic variation for example between an Indian and an English person as compared to two English people, in fact the evidence out there at the moment suggests that there's greater genetic variation between say two Scots than there would be between a Scottish person and someone of Indian origin. So we would say in terms of genetic variation it doesn't matter where you come from and that, it shouldn't really, with doing it proportionate to the numbers in the population to make it fair as opposed to for any genetic kind of justification for that, it's not that we're trying to find particular genes in particular populations.
>
> (PGD 9)

This interviewee rejected the idea that ethnic minorities were being included on the grounds that they represented genetically-defined groups. He referred to the common argument made by many geneticists that genetic variation is greater within than between such populations. We might therefore infer that this scientist conceived of ethnic minorities as *social* groups which do not necessarily reflect genetic variation, as he noted there is no 'genetic kind of justification' in the recruitment policy. He characterised UK Biobank as an 'inclusive' enterprise, and implied by his use of the term 'fair', that it is one that is perhaps operating with a consideration to social justice and equity.

These themes were elaborated upon by other interviewees whose discussions about the inclusion of ethnic minorities were cast in terms of it being a matter of participatory citizenship. One, borrowing the language of entitlements associated with democratic civil societies, remarked that the provision of recruitment materials in different languages would ensure that 'no one is disenfranchised as such from participation' (PGD 4). People are then enfranchised through the availability of information through appropriate languages that at least removes the linguistic barriers to their potential 'participation'.

The inclusion of all ethnic groups in UK Biobank was also seen in terms of the fact that two of the UK Biobank partners 'Are funded by tax payers who are paying for it, everyone who's paying their taxes, they should be able to get the benefit from it as well' (PGD 9). The right to be able to 'participate' in UK Biobank – and for the groups to which people might identify to potentially benefit from this research (a point onto which I come later) – stemmed from the fact that people from ethnic minority backgrounds are also tax-payers and should be included in a project supported by public funds.

The numbers

One interviewee discussed in depth the debate about how to include ethnic minorities in UK Biobank and in what numbers. He noted that there were two opposing positions and that UK Biobank was making a compromise between the two extremes. On the one hand, he discussed how some scientists had argued for Biobank not to include any subjects from non-white ethnic groups because 'we won't get a large enough sample size to answer anything scientifically of value' (PGD 15). On the other, drawing on the stance taken by Francis Collins, the head of NIH, on its plans for a national biobank study, there was the view that Biobank should over-sample from minorities. To go the first route, the interviewee argued was not ethical: 'I just think it's totally unacceptable from an ethical point of view, to exclude a substantial proportion of the population from this national study; it just seems to me to be totally unacceptable' (PGD 15).

However, taking the latter option was not a viable one for the UK because the size of some minority groups would mean that scientists on UK Biobank would have to recruit around 100,000 people, and achieve recruitment rates of over 50 per cent for many of these groups. Therefore, the compromise was to settle for something in between these two positions. As the interviewee explained:

> Is there a scientific argument for recruiting some people, but less than 100,000? And this is where the argument comes up about a universal

> control series. At the moment the UK is trying to develop a common control series so that's a series of people who've basically been genotyped so that you can get hold of the gene frequencies in them but that they are a representative sample, they don't have any particular disease and so you can then compare the gene frequencies in them to any other case series that might come along. [. . .] So Biobank will be able to provide a fantastic resource, which is, not replicated anywhere else in the UK of people from ethnic minorities that can be there compared to case groups for ethnic minorities.
>
> (PGD 15)

Therefore, UK Biobank, while not providing the means to carry out a gene–environment interaction study for each ethnic group would at least provide the ability for other researchers to undertake case-control studies with the information and samples collected from the groups that would be included. This solution for UK Biobank helped to secure its political and scientific credibility by meeting the political need to include ethnic minorities and the scientific one that their inclusion is in sufficient numbers to be of some scientific value.

Benefits for ethnic minorities from being in UK Biobank

Following on from this, interviewees were asked about what they considered to be the potential benefits for ethnic minorities from being included in Biobank. Would the project assist with elucidating and maybe redressing disparities amongst ethnic groups in relation, for example, to disease incidence and mortality? One interviewee reflected on the kinds of resources that UK Biobank would generate:

> What's quite likely is that certain factors like ethnicity will help to create meaningful cohorts that other people can capitalise upon and that's the first and best thing and that will be a major advance for the UK really which typically has got under-researched ethnic groups.
>
> (PGD 4)

Specifically in relation to the creation of the case-controls, the Biobank would provide a set of valuable research resources that had not previously been available in the UK. Therefore, it would help to at least redress the history of under-researching ethnic minority groups. The benefits that could flow from these resources were highlighted by another interviewee, who said:

> I think potentially it could have quite a lot of benefits because if you can perhaps predict which population is going to get which kinds of diseases and so on it could be very useful for the Health Service and

> so on, I think. Also certain groups of the population are more susceptible to certain kinds of diseases than others, and again thinking of diabetes in South Asians.
>
> (PGD 14)

However, one interviewee did sound a note of caution about the possible implications of uncovering health information on specific ethnic groups. He remarked that:

> Within the UK, once we start getting results from ethnic minorities then there's effectively a conflict [. . .] first of all the results that you find might imply that something very important for that group, which would mean that they really, you might want to intervene differently in that group which obviously would be to their benefit so you can't ignore those results. On the other hand those results can then be used to discriminate against those groups, either in terms of an additional stigma when, when there's a group that's already being stigmatised anyway.
>
> (PGD 15)

On the one hand, such data could help researchers and clinicians to tailor treatments, for example, so that they are more effective for certain groups. However, there was also a concern that such information could also reinforce negative perceptions of these groups and exacerbate existing stigma.

Category-making and identity

In addition to including ethnic minorities in UK Biobank in ways that would seem equitable, scientifically meaningful and practicable, the scientists at the Biobank were also concerned with issues around classification. They decided to adopt the ethnic group classification produced as part of the UK National Census in 2001. As an interviewee explained:

> At the moment we're proposing to use the Census based tool which is a self assigned ethnicity categorisation [. . .] I guess in terms of using those categories and arguing for it there are very few classifications that [. . .] we've come across that have been developed with community engagement and the advantage of a community engaged classification that you can say has been used and does appear to be acceptable to community groups.
>
> (PGD 4)

As argued elsewhere (Smart *et al.* in press), the UK Census provides a classification that is regarded as widely acceptable to the people asked to assign themselves to one of its categories. The reference to 'community groups' by this interviewee would appear to mean 'minority groups' who, he supposes, might have particular objections to a classification that did not include them or identified them through an inappropriate category. The same interviewee noted the importance of using the right categories in research:

> It seems [. . .] that communities in the UK actually probably prefer to be called Black Caribbean as opposed to African Caribbean but it changes in the approach to calling people, or describing people in a certain way. For example, the unacceptability of terms like coloured which is you know got no, it's got no specific definition of the community, it's got no identity to it, it's simply a broad-brush generalisation.
>
> (PGD 4)

The suggestion that 'coloured' as a generalised category is one without 'identity' is a telling comment. It underscores the importance of researchers using categories with which potential subjects identify. The failure to do so could cause offence or pose other measurement problems for them. Therefore, the Census was chosen by UK Biobank because its classification had been shown to work in practice in 2001, and indeed had been subject to various public consultations before its adoption by the National Office of Statistics (Bosveld *et al.* 2002).

Discussion

These interviews indicate how those involved in UK Biobank have been keen to stress that it is an inclusive enterprise with the potential to benefit all sections of society, and have engaged with the considerable scientific and practical issues involved to actively include minority groups in a meaningful way. The evidence suggests that researchers at UK Biobank were reflexive about past recruitment practices that led to the under-representation or exclusion of minorities from scientific and medical studies. They constructed the inclusion of minorities on grounds of social justice and invoked the language of citizenship in doing so. One interviewee was careful to say that ethnic minorities were not considered to be genetically-meaningful groups but this was undermined perhaps by the fact that ethnic minority subjects will be genotyped to act as a resource for case-control studies. So, while their inclusion might be framed in social terms, their value in the Biobank will lie primarily in their genetic data.

The representation of ethnic minorities was also a highly significant and somewhat contested issue about what would constitute meaningful numbers for different kinds of studies. My conclusion is that this decision involved a mixture of scientific and pragmatic considerations on the part of the Biobank. Given the history of under-representation, there was also a hope that UK Biobank would be beneficial for minority groups and would address their health needs. However, concerns about the dangers of stigmatisation were voiced by one interviewee. Finally, the choice of classification for UK Biobank – the National Census ethnic group classification – was chosen because it was seen to be accepted by minority groups, used appropriate categories and was inclusive of all groups. One criticism levelled at UK Biobank is that debates about ethics and governance have effectively been compartmentalised from scientific issues (Wallace 2005). However, my discussion of the interviews reveals how in their accounts of the inclusion and study of ethnic minorities in UK Biobank, switched back and forth between scientific, pragmatic and social-political considerations.

Conclusion

In this chapter I have argued that one of the key challenges that faced UK Biobank was to co-produce the scientific and socio-political orders of 'inclusiveness'. On the one hand, there were scientific and practical concerns about achieving sufficient numbers from ethnic groups so that the biobank would be suitably powered to answer its scientific questions. On the other, was the need for the project to be inclusive, open to all people to volunteer and one which would not marginalise minorities many of whom of course suffer from continuing health inequalities. In the end, through the co-production of the natural and social orders, UK Biobank has emerged as a compromise of scientific, pragmatic and socio-political considerations. Whether this compromise will prove to be a successful one remains to be seen. Given that UK Biobank only began recruitment on a large-scale towards the end of 2006, it is too early to say whether the scientists will be successful in generating their target numbers of volunteers from their specified ethnic minority groups, or whether the resources that they aim to generate will be of benefit to those groups in the way that they currently hope.

In relation to the social science literature on biobanks, by using the co-production as its framework, this chapter has aimed to show that the legitimacy of initiatives such as biobanks can be understood as being established in the interaction of scientific knowledge production and the practices and norms of governance. Social science analysis in this area would benefit from such an approach that simultaneously seeks to understand legitimacy in scientific and social-political terms. In addition, this chapter has hopefully brought to the fore a relatively under-researched set of issues related to the inclusion and representation of ethnic minorities in biobanks. While my analysis has focused mainly on the UK context, how the issue of ethnicity and ethnic group differences are recognised and conceptualised within various biobank initiatives worldwide arguably deserves further attention.

Acknowledgements

This work was supported by the Wellcome Trust Biomedical Ethics Programme, grant number: 073524/Z/03/Z/AW/HH (Project Team: Paul Martin,

Richard Ashcroft, George Ellison, Andrew Smart and Richard Tutton). For their comments on earlier drafts of this chapter I would like to thank George Ellison, Andrew Smart, Pritti Mehta, the editors Alan and Herbert, as well as my colleagues at the Institute for Science and Society. I would also like to acknowledge the researchers who agreed to be interviewed for the study.

Notes

1 From 1932 to 1972, the Tuskegee syphilis study in Alabama, USA, involved around 600 African-American men 'in what is now considered one of the worst examples of arrogance, racism, and duplicity in American medical research' (Reverby 2000). Publicity about this study – and other studies that also involved black subjects – fuelled a number of important changes in research governance (Rothman 2003).

2 See, for example, Arnason and Simpson's (2003) discussion of the conflict over deCODE and its plans to establish the Health Sector Database in Iceland. They suggest that it centred on what it meant to be Icelandic, locating both the source and the essence of '"Icelandicness' in the very building blocks of people's bodies and offers novel ways of contemplating and reasserting identity' (Arnason and Simpson 2003: 549). They whose bodies and histories would count as 'Icelandic', would the Icelandic Biogenetic Project include the so-called 'latecomers'? As Rose (2001) noted, the emphasis on the genetic and physical similarity of the population overlooked the minorities who live in Iceland, among them people with origins in India, Eastern Europe, the Balkans and Vietnam.

3 Despite such efforts, evidence suggests that ethnic minorities are continuing to be under-represented or excluded from studies, especially those based in Europe (Ranganathan and Bhopal 2006) and that often researchers fail to provide clear explanations of why ethnic minorities are missing from studies (Hussain-Gambles 2003).

4 It is also important to acknowledge that this political construct of multiculturalism has been recently contested, especially in light of the London bombings of 7 July 2005 (BBC News 2005).

5 The other national biobank in the UK is 'Generation Scotland', formed from a partnership between the Scottish Executive and Scottish universities' medical schools. This smaller project will recruit 50,000 subjects and family members from Scotland only. There are a number of interesting scientific and organisational differences between these two initiatives, not least in relation to the prominence that appears to have been given to the inclusion of ethnic minorities in their cohorts. Available evidence suggests that Generation Scotland has adopted the policy of 'welcoming' the involvement of all ethnic and cultural groups but has not adopted any specific set of policies to ensure the inclusion of minorities in certain numbers (Generation Scotland 2006).

6 This quote comes from an interview conducted with one of the UK Biobank scientists.

References

Arnason, A. and Simpson, B. (2003) 'Refractions through culture: the new genomics in Iceland', *Ethnos*, 68 (4): 533–53.

BBC News Online (2005) 'How multicultural is Britain?', available online at: http://news.bbc.co.uk/1/hi/talking_point/4741753.stm (accessed 31 July 2006).

Bosveld, K., Connolly, H. and Rendall, M. (2002) 'A guide to comparing 1991 and 2001 census ethnic group data', available online at www.statistics.gov.uk/articles/nojournal/GuideV9.pdf (accessed 31 July 2006).

British Embassy in the United States (2006) 'Multicultural Britain', available online at www.britainusa.com/index.asp (accessed 31 July 2006).

Busby, H. and Martin, P. (in press) 'Biobanks and imagined communities', *Science as Culture*.

Carson, P., Ziesche, S., Johnson, G. and Cohn, J. N. (1999) 'Racial differences in response to therapy for heart failure: analysis of the vasodilator-heart failure trials', *Journal of Cardiac Failure*, 5 (3): 178–87.

Chadwick, R. (2003) 'Genomics, public health and identity', *Acta Bioethica*, 9: 209–18.

Collins, F. (2004) 'The case for a US prospective cohort study of genes and environment', *Nature*, 429: 475–7.

Duster, T. (2005) 'Race and reification in science', *Science*, 307: 1050–1.

Epstein, S. (2004) 'Bodily differences and collective identities: the politics of gender and race in biomedical research in the United States', *Body and Society*, 10, 183–203.

Epstein, S. (2007) *Inclusion: The Politics of Difference in Medical Research*, Chicago, IL: University of Chicago Press.

Friedman, D. J., Cohen, B. B., Averback, A. R. and Norton, J. M. (2000) 'Race/ethnicity and OMB directive 15: implications for state public health practice', *American Journal of Public Health*, 90: 1714–19.

Gambles-Hussain, Mavhash (2003) 'Ethnic minority under-representation in clinical trials: whose responsibility is it anyway?', *Journal of Health Organisation and Management*, 17: 138–43.

Generation Scotland (2006) Available online at www.generationscotland.com (accessed 12 June 2006).

Human Genetics Commission (2002) *Inside Information: Balancing Interests in the Use of Personal Genetic Data*. London: Department of Health.

Jasanoff, Sheila (2004) 'The idiom of co-production', in S. Jasanoff (ed.), *States of Knowledge: The Co-production of the Natural and Social Order*. London and New York: Routledge.

Juengst, E. T. (1998) 'Group identity and human diversity: keeping biology straight from culture', *American Journal of Human Genetics*, 63: 673–7.

Kahn, J. (2004) 'How a drug becomes "ethnic": law, commerce, and the production of racial categories in medicine', *Yale Journal of Health Policy, Law and Ethics*, IV, 1–46.

Kaye, J. (2004) 'Abandoning informed consent: the case of genetic research in population collections', in R. Tutton and O. Corrigan (eds), *Genetic Databases: Socio-ethical Issues in the Collection and Use of DNA*. London and New York: Routledge.

Lewis, G. (2004) 'Tissue collection and the pharmaceutical industry: investigating corporate biobanks', in R. Tutton and O. Corrigan (eds), *Genetic Databases: Socio-ethical Issues in the Collection and Use of DNA*. London and New York: Routledge.

Martin, P. (2001) 'Genetic governance: the risks, oversight and regulation of genetic databases in the UK', *New Genetics and Society*, 20: 157–84.

Mehta, P. and Saggar, A. (2005) 'Ethnicity, equity and access to genetic services – the UK perspective', *Annals of Human Biology*, 32: 204–10.

NIH (2006) 'Policy issues associated with undertaking a large US population cohort project on genes, environment, and disease', draft report of the Secretary's Advisory Committee on Genetics, Health, and Society, May 2006, available online at www.nih.gov (accessed 31 July 2006).

NIH (National Institutes of Health) (2001) 'Policy and guidelines on the inclusion of women and minorities as subjects in clinical research – amended, October 2001', available online at www.nih.gov (accessed 31 July 2006).

Palsson, G. (2002) 'The life of family trees and the book of Icelanders', *Medical Anthropology*, 21: 337–67.

Petersen, A. (2006) Securing our genetic health: engendering trust in UK Biobank, *Sociology of Health and Illness*, 27: 271–92.

Ranganathan, M. and Bhopal, R. (2006) 'Exclusion and inclusion of nonwhite ethnic minority groups in 72 North American and European cardiovascular cohort studies', *PLoS Medicine*, 3: 329–36.

Rathore, S. S. and Krumholz, H. M. (2003) 'Race, ethnic group, and clinical research', *BMJ,* 327, 763–4.

Reardon, J (2005). *Race to the Finish: Identity and Governance in the Age of Genomics*, Princeton, NJ, and Oxford: Princeton University Press.

Reardon, J. (2001) 'The Human Genome Diversity Project: a failure to co-produce natural and social order', *Social Studies of Science*, 31: 357–88.

Reverby, S. (ed.) (2000) *Tuskegee's Truths: Rethinking the Tuskegee Syphilis Study*, London: University of North Carolina Press.

Rochon, P. A., Berger, P. B. and Gordon, M. (1998) 'The evolution of clinical trials: inclusion and representation', *Canadian Medical Association Journal*, 159: 1373–4.

Rose, Hilary (2001) *The Commodification of Bioinformation: The Icelandic Health Sector Database*. London: Wellcome Trust.

Rothman, D. (2003) *Strangers at the Bedside: A History of How Law and Bioethics Transformed Medical Decision Making*. New York: Basic Books.

Salter, B. and Jones, M. (2005) 'Biobanks and bioethics: the politics of legitimation', *Journal of European Public Policy*, 12 (4): 710–34.

Smart, A., Tutton, R., Ellison, G. T. H., Martin, P. and Ashcroft, R. (in press) 'The standardization of race and ethnicity in journal editorials and UK Biobanks', *Social Studies of Science*.

Tate, S. and Goldstein, D. B. (2004) 'Will tomorrow's medicines work for everyone?', *Nature Genetics*, 36: S34–S4.

UK Biobank (2002a) *A Protocol for the UK Biobank: A Study of Genes, Environment and Health*. London: MRC, Wellcome Trust, and Department of Health. Available online at www.ukbiobank.ac.uk/documents/draft_protocol.pdf. (accessed 31 July 2006).

UK Biobank (2002b) *Peer Review of Protocol for Biobank UK*. London: MRC, Wellcome Trust and Department of Health. Available online at www.mrc.ac.uk/strategy-biobank (accessed 31 July 2006).

UK Biobank Ethics and Governance Framework, Version 1.0 (2003). London: MRC, Wellcome Trust and Department of Health. Available online at: www.ukbiobank.ac.uk/documents/egf-comment-version.doc. (accessed 31 July 2006).

UK Biobank Ethics and Governance Framework: Summary of Comments on Version 1.0 (2004). London: MRC, Wellcome Trust and Department of Health. Available online at www.ukbiobank.ac.uk/ethics/efg.php (accessed 31 July 2006).

Wallace, H. (2005) 'The development of UK Biobank: excluding scientific controversy from ethical debate', *Critical Public Health*, 15, 323–33.

11 The informed consenters

Governing biobanks in Scandinavia

Lars Øystein Ursin, Klaus Hoeyer and John-Arne Skolbekken

In the three Scandinavian countries, Norway, Sweden and Denmark, debate about the regulation of biobank research is surrounded by a classical ambiguity: on the one hand, population-based genetic research is seen as a source of medical hope (hence research should not be impeded), on the other hand, it is presented as a potential threat to existing institutions, norms and values (cf. Mulkay 1993). Though revolving around similar hopes and fears, however, the regulatory responses have played out differently in the three countries.

In this chapter, we explore such differences as they can be identified in white papers, acts and circulars, and relate them to the views and perceptions of researchers and research subjects in biobank studies. The views of the researchers and research participants are surprisingly similar across national contexts, just as they stand in remarkably similar contrast to the slightly different concerns demonstrated in the various legal efforts. If the legislative efforts do not reflect the concerns of the donating citizens, nor of the researchers using biobanks, how are we to understand the object of the regulatory effort? We suggest that the differences between countries obscure more fundamental governmental similarities. We argue that the legal approaches can be seen to share particular governmental notions, notably in the way the approaches *enact* a particular, paradoxical, understanding of the role of the citizen in public policy: an ideal of autonomy is fused with a notion of education of the masses; a search for the uncontaminated and independent citizen as arbiter of moral judgment is fused with strong efforts to enrol this 'subject' in public health policy-making.

Biobanking as a challenge for governance

In all three countries the citizens seem to have been released from the moral obligation to contribute to the research facilitated by the institutions of the welfare state. It is no longer taken for granted that people are willing to assist the progress of research. Biobanking becomes a touch point of ambiguous science and technology, and of ambiguous rights and reflections of the citizen. Biobank regulation in a sense exemplifies and institutionalizes reflexive modernity in the sense of Anthony Giddens: 'Today scientific findings and technological change impact upon us in an immediate way – we have more dialogic or interrogatory relation with them than in the past. We are all finding out what philosophers of science have uncovered, that science rests upon organized scepticism' (Giddens, in Pierson 1998: 117). Legislators on the one hand set participants free, but on the other the regulation can be seen to contribute to the shaping of the thinking of the citizens. It does not just protect them from state abuse: 'One must abandon the conventional ways of ascribing ethical value to the opposition between subject and object, in which subjectivity is privileged as the authentic and natural locus of moral autonomy: we are governed as much through subjectification as through objectification' (Rose 1999). What happens when the state simultaneously withdraws from and frames decision-making in the health sector (Baumann 1995)?

It is important to note how the regulatory reasoning in the three countries exhibits a particular rhetorical framing of the legal effort as a matter of *ethics* (cf. Petersen 2005a): reference is made to historical instances of medical atrocities, a paternalist tradition in healthcare and scenarios highlighting future genetic discrimination. Central to the framing of the regulatory issue as a matter of ethics is the role played by the practice of informed consent. Here we set aside questions of how or whether different practices of informed consent are justified or whether they reach their proclaimed aims. Instead, we address the reflections of research participants, researchers and policy makers on the intervention of epidemiology in the private sphere of the individual. These reflections make us ask which type of regulation informed consent and other features of biobank law seem to represent. Doing this, we want to combine an analysis of the way biobank participation is regulated and the way the participants are governed: we both look at the legislative attempts to liberate the participants, and how these solutions thereby reconstruct the identity of participants, and conduct their conduct (Foucault 1979).

The practice and discussion of informed consent highlights the relation between healthcare and the research participant, the state and the citizen, the individual and the culture. On the one hand, biobanking brings forth questions concerning individual rights versus collective interests. Biobank regulation becomes an arena for negotiation of the state/citizen relation. How does biobanking intervene into the privacy of the participants? What are the dangers and benefits perceived in this intervention? Which individual and collective interests are recognized? Which ideals are at play? On the other hand, biobanking gives reason to consider questions beyond the state/citizen relation. The regulatory framing of biobank research as an ethical issue might have contributed to restricting academic interest in this relationship. In a sense, social science inquiry has mimicked the ethical mode of analysis in which 'the interests of society are balanced against the interests of the individual'. Subsequently, questions concerning other types of regulation of biobank research in terms of funding, academic freedom, property regimes, incentive structures and public/private partnerships are left unasked (Petersen 2003). Ironically, the latter series of issues is at the heart of the concerns of the donating public, and this chapter explores the ethical mode of policy-making as a way of avoiding their confrontation.

Epidemiology in the Scandinavian welfare states

Scandinavia is known for a proud epidemiological tradition based on extensive registries, systematic clinical records, and impressive biological repositories. Though human biological material has been collected systematically for decades in hospitals and medical research institutions, it was only during the late 1990s that they, by then known as biobanks, became an object of political controversy and regulatory interventions. A conservative estimate is that there are 20 million samples stored in Denmark (5 million inhabitants), 60 million samples stored in Sweden (8½ million inhabitants) and 40 million samples stored in Norway (4½ million inhabitants). In Denmark alone, more than 900,000 pathological samples are registered in the central pathological registry every year (Vyberg *et al.* 2005; Patologidatabank 2006). This activity is considerably more intensive than in, for example, the United States where a Rand Corporation report in 2000 estimated storage of 307 million samples in the whole country including blood and organ banks (RAND 2000). This has for instance led the National Institute of Health in the US to finance research projects based on Scandinavian biobank material. From an epidemiological perspective, Scandinavia is recognizable on the world map.

It is sometimes argued that the potential of Scandinavian biobanks should oblige Scandinavians to pursue biobank research for the sake of the global community. As Ole Jan Iversen at the Functional Genomics Programme (FUGE) of the Norwegian Research Council puts it: 'In Norway we have a well-organized health care system, large epidemiological studies and biobanks, and the necessary economic means. Our society represents the ideal starting point for genetic epidemiology. Our national advantages commit us globally'[1] (Fuge 2006).

Due to the scale of the biobank activities even small organizational amendments potentially imply considerable costs and affect a great number of people. Commenting on the first draft of the new Human Tissue Bill in the UK, Furness and Sullivan (2004) estimated that the proposed strengthening of the consent requirement would cost one minute per sample which would amount to 1,339 full time jobs or approximately the same as the entire staff of a medium-sized National Health Service (NHS) hospital. Changes in the handling of human tissue samples are in no way insignificant to the health services.

The health services in the Scandinavian countries are so-called Beveridge systems, named after Lord Beveridge, the architect behind the British NHS founded in 1946. This type of healthcare system is characterized by universal access for all inhabitants and makes the health services a key object of political deliberation in a way different from, for example, the United States where the majority of health expenditure is privately financed. Every Scandinavian citizen has a personal identity number used in practically all affairs relating to the public authorities. Internally, the healthcare systems are organized differently in the three countries. Both in Norway and Denmark major reforms have recently reorganized the assignment of tasks between national, regional and local levels of government. The mode of politics, however, share some features known as 'the Scandinavian model', which is in contrast to conventional democracy models such as the Anglo-Saxon Westminster model or the Continental consensus model (Lane and Ersson 1996). The institutions of the Scandinavian model typically express compromise politics within an overall constitutional frame of the Westminster type. Government is based on multiparty systems, and there is an extensive tradition for prolonged hearings of draft bills among a wide set of organizations. Moreover, Denmark is renowned for having developed a particular deliberation model in which ordinary citizens are selected to form a citizen's panel, which consults a number of experts and on the basis of this formulates recommendations to Parliament. This type of public consultation has been praised also by scholars in the tradition of Science and Technology Studies (STS) as a prototype of a more democratic approach to politics in the field of science and technology, and is now being copied in a number of countries with interesting implications (Irwin 2004; Irwin 2006). The Swedish system is tied more closely to selected groups of experts, but deliberate attempts of involving the ordinary citizens in science and technology have multiplied in all three countries in recent years.

In developing a national policy stance on the biobank issue the Scandinavian model implied an extensive process of consultations with a great number of experts and organizations. However, the course of events has developed differently in the three countries and the resulting pieces of legislation differ significantly, at least on the surface. Norway and Sweden have created discrete biobank acts, while the relevant Danish ministry initiated only a circular. Correspondingly, Swedish and Norwegian changes of practice seem more encompassing than the Danish amendments.

A comparison of the Scandinavian legislation on biobanks

The Norwegian Biobank Act requires explicit, informed and written informed consent for all storage of human tissue for research purposes. The Swedish Biobank Act similarly requires written informed consent for storage of human tissue for more than two months, though allowing the ethics committees to waive the consent requirements in named instances. The Danish circular, on the other hand, establishes an opt-out system.

In Norway and Sweden, some specification and procedures were added through the acts concerning registration and standardization of the various biobanks constructed in relation to public health services. In Sweden, biobanks in the private sector were left unaffected by the new law, i.e. the collection of the tissue by the pharmaceutical industry was not addressed by parliament. The Danish circular implied establishing a new registry for people who would restrict usage of their tissue to their own treatment and diagnostic purposes, thus avoiding further confrontation with requests from researchers wanting to use their samples. Researchers wanting to use stored tissue samples must first check with the registry to make sure that none of their potential participants have declined participation. The underlying premise for the circular was that existing law should regulate both biobank administration and the consent issue. However, existing law in this field is ambiguous. Though a number of rules concerning registration of biobanks probably were more elaborate in Denmark than in Norway and Sweden, three different acts have relevance for the consent issue with different wordings in each. Taken together, informed consent is mandatory when tissue is taken explicitly for research purposes (not for storage only) and when stored tissue is used for research implying health risks, menace for, or stigmatization of the individual. Research Ethics Committees (REC), which are the Scandinavian equivalent to the American Institutional Review Boards (IRB), should decide when such risks are relevant, in fact making it their task to administer the consent requirement. Traditionally, biobank research has not been seen as posing risks to the individual. Researchers would, therefore, in effect be free to use stored samples unless the donors are registered in the new registry. Some lawyers, however, argue that this interpretation conflicts with Article 22 of the Council of Europe's Convention on Human Rights and Biomedicine where informed consent is made mandatory for all research projects involving human tissue (Hartlev 2005: 469).

The negotiations on the issue never caught the attention of the public in Denmark in the same way as in Sweden and Norway (Vedel 2004), let alone Iceland (Pálsson and Harðardóttir 2002). In Sweden the issue of biobanking was launched by a popular tabloid newspaper focusing on the use of stored tissue as 'yet another infringement' of the state on the rights of the individual (see Hoeyer 2005, for an analysis of the naming and framing of human tissue as a legal problem in Sweden). This framing helps explain why legislation was confined to tissue collected in connection with the public health services. Furthermore, it might be important that commercial initiatives inspired by the Icelandic company deCODE Genetics were considered in both Norway and Sweden – in relation to biological material collected by the authorities. The boundary between commerce, public responsibility and individual rights were thus more clearly at stake in Norway and Sweden than in Denmark.

The required specific consent for participation in research biobanks in Norway has met massive critique from researchers, which deem the current Biobank Act unethical in impeding important research. In an effort to reduce regulatory impediments to medical research (which at present is affected by more than twenty laws), a single law regulating this activity has been proposed (NOU 2005). If this proposal is passed as a law, it will affect biobank research through a redefinition of biobanks and an acceptance of broad consent from donors. At the time of writing, the hearing process is coming to an end, indicating that a number of important public institutions and private organizations have taken a rather critical stance to the proposal. How this will influence the law-making process remains to be seen.

In Sweden health professionals have complained that without storage of tissue used for diagnostic purposes, they cannot document that they have taken the proper action in concrete cases of patient treatment. It is often necessary to go back to a sample and see if the tests were correct. Storage may also serve as proof in case of patient complaints. But Swedish regulation aims to protect the rights of the people that do not share the stated aims of the welfare state: research, industrial development, improved diagnostics and quality control measures. The rights of the patients wanting to ensure that every sample is only used for their own benefit are pre-eminent. In Denmark, the legal action is likewise aimed at ensuring the rights of the people who want to refrain from any type of research participation – regardless of the risks posed to themselves or benefits to others.

Having identified these differences between the legal paths taken in the three countries, we now wish to explore what simultaneously unites them. The regulatory reasoning surrounding biobanks in Scandinavia has revolved around balancing the rights of the individual and the interests of the common good. This is a well-known theme in medical ethics. In Denmark the common good has been perceived less at odds with the rights of the individual than in Sweden and Norway, and a more relaxed approach to the consent issue seems to be the result. However, the legislative effort in all three cases concentrates on *securing the individual the right to decline supporting public research*. The legislation claims to aim at the protection of the individual. Since participation in a biobank is restricted to the use of tissue already procured from medical treatment, or minimally invasive procedures like giving a blood sample, bodily harm cannot be the primary concern. In fact, it can be seen as an open question what the primary concern is. Also we need to consider why patients are offered the possibility of retaining storage of tissue, free of charge, *only* for their own diagnosis and treatment? Why is consent suddenly important for participation in research?

Defining and protecting privacy

It appears that the regulation of biobanks is a part of the general protection of the privacy of citizens. The public authorities, who are granted extensive measures for control (e.g. personal identification numbers) and imbued with great expectations in relation to healthcare delivery, are eager to signal respect for the autonomy of the individual citizen. The human biological repositories come to represent an apt metaphorical showground where authorities can signal proper conduct in relation to the precarious boundary between public and private. What makes biobank information (genetic plus phenotypic information) distinctive is the fact that in this case the body is a source of information. The bodily origin of part of this information makes the stored material doubly potential: it concerns the biological dispositions of a person, and it is written in a code which we are learning to read. At the time of legislation, human genetics seemed to promise a dramatic increase in our ability to bring out the information potential of biological material in the years to come (Petersen 2001). When it comes to genetic information nobody knows its content: accordingly, the handling of it might be harmful for the individual – or the relation between the individual and the state – in some way or other. The legislation is consequently clearly influenced by the precautionary principle.

But what is the relation between biobank information and the person: where does the individual end (Tutton 2004)? The white paper (NOU) preceding the current Norwegian Biobank Act defines biobank material as a part of the individual: 'Cells, tissue and other biological material in biobanks must be included to respect human dignity, and is not to be viewed as disconnected from the donors' (NOU 2001: 19; 5.1.9). The message seems to be that both the state and the individual should learn to treat biobank material with care and respect, as a proxy for the individual. The biological information is private in the sense that it represents both the identity and the bodily integrity of the citizens. Consequently, just as you respect someone's private sphere by asking for permission if you want to enter, the definition offered by the white paper naturally leads to the consent requirement of the Norwegian biobank Act: '. . . collection, storage and use of human biological material for research purposes [requires] an explicit and informed consent from the donor. The same is the case for storage and use of information linked to the biological material' (The Norwegian Biobank Act, ch.3; §12).

The stored sample stands in a metonymical relationship with the individual, and the management of the biobank comes to epitomize the governmental approach to the citizens in its care. Here the public health system is central to understand the negotiation of trust: the public authorities are responsible agents for the welfare of the individual in a very intimate sense. Interestingly, however, the obligation of the individual to contribute to the maintenance of a research-based healthcare system is set aside by the three pieces of law emphasizing only the rights of the individual.

Informed consent and the politics of ethics

Mimicking known themes from medical ethics, the legislative work in the Scandinavian countries sets itself against the background of the history of eugenics, unethical medical experiments during and after the WWII, a history of paternalistic thinking in healthcare, and position itself in relation to documents of research ethics like the Nuremberg Code and the Helsinki Declaration. Thus, contemporary biobanking is contrasted with a past in which both science and government is viewed as lacking the respect for autonomy that permeates medicine today (Koch 2004). Accordingly, it becomes important for the legislative initiatives to indicate an anti-paternalist impetus.

Once the political field is framed as a matter of ethics, the themes named in the debates come to reflect well-known ethical issues. For example, whether genetics potentially opens the door to genetic discrimination of individuals and groups, and whether this might imply a disruption of our ideals of a 'human and solidary society' (NOU 2001;19; 5.1.4), and whether the autonomy of the individual is sacrificed for the sake of societal ends. It becomes necessary for an act on biobanks to explicitly embody the ideal of a diversified and inclusive society. The legal regulation of biobank research should state its aim to respect the uniqueness of every citizen, by protecting both their genetic and social privacy (Helgesson 2003).

In Scandinavian law, this is accomplished through the informed consent requirements. The Norwegian white paper mentioned above explains the consent requirements of the proposed Act in this way:

> The norm must be that collection, storage and handling is based on an informed, voluntary and individual consent. Biobank research can create different forms of dependency, for instance between patient groups with rare diseases, and research communities which are dependent on information from these. The promotion of public health and industry can also put pressure on the research. The committee finds that these kinds of pressure can be controlled only if the individual freely, and at any time, can withdraw their consent.
>
> (NOU 2001: 5.1.3)

The tool to protect the individual from harm turns out to be a tool that also can be used to avoid research being perverted by motives alien to the common good of the society. To make the consent informed, research designs have to be publicly accessible, and shield the participants from harmful research, and the society from unethical research. Every participant should know and approve of the research s/he takes part in – if not, they are free to leave. To recruit and keep voluntary research participants, research projects have to build trust. In this way, the informed consent requirement can be claimed to serve three different purposes: to protect participants from pressure to take part in harmful research, to protect researchers from pressure to do research led by shady motives, and to ensure that biobank research is open to public scrutiny, control and debate. The threefold basis of the consent requirement – the promotion of the privacy, autonomy and safety of the individual – is here nicely summed up as the ethical duty to respect every individual as an end in her/himself:

> Information and consent requirements are the foundation for safety and trust in health care and medical research in general, and in biobank enterprise as a part of medical diagnostics, treatment, and research in particular. (NOU 2001; 19; 5.1.3) . . . most people have little information about and insight into the daily activity at hospitals, research institutes, and in industrial and commercial development. This fact can give rise to suspicion and resistance. The requests of the researchers and commercial companies can lead to secrecy. If research and industrial and commercial development is not subject to general ethical evaluations, this might undermine the basic principle of treating human beings as an end in themselves, and not only as a means.
>
> (NOU 2001: 19; 5.1.7)

The emphasis on the *lack* of information and insight is important: the consent procedure is not just about delegating decision-making power. The *consenter* has to be *informed*. This implies figuring out what the population needs to know – and accordingly most of the battles surrounding the consent requirement revolves around figuring out what to write on the information and invitation sheets.

Participants' views

As the blood and other forms of tissue stored in the various hospitals, laboratories and healthcare centres of the Scandinavian welfare states became a hot ethico-political issue, funding was established for a number of studies of the ethical, legal and social aspects of the previously somewhat marginal activity of biobanking. Suddenly, a number of studies were funded to elicit the views of the donating public. The authors of this chapter have benefited from this fact. As remarked by Leigh Turner, 'if you are an anthropologist or bioethicist, government-funded genethics research programs are the wealthiest 'sugar daddies' you are likely to find' (Turner 2003: 1282). In the following we draw accordingly on our own and other people's work to elicit themes in the reasoning of participants in biobank research. The perspective of the participants on biobank research allows us to discover some puzzling contrasts between the donating citizens and the concerns of the policymakers explained above.

Members of the bioethics research group at the Norwegian University of Technology and Science (NTNU) have conducted a focus group study comprising participants in the Nord-Trøndelag Health Study (HUNT) biobank, biobank researchers and associates, as well as former participants which had withdrawn their consent in 2002 (Skolbekken *et al.* 2005). In 1995–7 blood samples and questionnaire data were collected from 65.000 donors in the county of Nord-Trøndelag situated in the middle of Norway. Together with questionnaire data collected from the same population in 1984–6, this material constitutes the HUNT study. Collection of data for a third study started in the fall of 2006 and will be completed in 2008. The focus groups consisted of an equal number of men and women, including participants both from rural and urban areas of the county. The discussion themes were: the use and abuse of biobank material, the decision to give consent, duty vs autonomy in biobank research, genetic vs other kinds of health information, and commercialization of biobank research.

The focus group participants were all rather vague when asked to state their hopes, fears and expectations concerning biobank research. Hopes and expectations were altruistically related to the development of improved therapies for the benefit of future generations. Fears were primarily related to a possible future commercialization of the research project, which might lead researchers to use controversial means in searching for knowledge, and aim for profitable areas in medicine rather than the alleviation of human suffering. The participants placed a great deal of trust in the researchers at HUNT, the REC and the Norwegian regulation of biobank research. They generally viewed the Norwegian society as well-regulated, well suited to exclude any abuse of personal (health) data. Even the possibilities of a linkage of registers were not particularly feared, as long as the research is not commercialized. The use of active broad consent was regarded as satisfactory, as long as information about the research pursued by HUNT is readily accessible by the public, and the participants in HUNT can opt out at any time. None of the participants in the focus group study thought that participation in HUNT should be mandatory, but some expressed the view that those who did not participate 'should search their conscience'.

Similar findings are reported in a study involving donors to Medical Biobank in the north of Sweden (Hoeyer 2003; Hoeyer 2005; Hoeyer and Lynöe 2006). Medical Biobank was initiated in 1985 and contains tissue samples and questionnaire data from 78,000 people and since 1999 has been at the disposal of a genomics company, UmanGenomics. In qualitative interviews with donors it was again obvious that people placed trust in the biobank research and the regulation of it, which was viewed as aiming for the common good. However, the participants did not think that they were in a position to assess the legitimacy of the biobank research. Therefore they considered the informed consent procedure more as a sign of being respected, than as their partaking of responsibility for the research. In a survey among donors to Medical Biobank, less than 4 per cent thought that the most important issue with respect to biobanks was whether they were personally informed (Hoeyer *et al.* 2005). Similar finding were found in surveys in the general public (Hoeyer *et al.* 2004; Kettis-Lindblad *et al.* 2005).

As in Norway, commercialization and the profit motive were generally articulated as threats to proper research during interviews; however, the fact that a private company would have access to the donated samples rarely seemed to influence donors. A survey among donors to another biobank in the same region confirmed this finding: only 2½ per cent of the participants were opposed to industrial genetic research on their samples (Stegmayr and

Asplund 2002). The dislike of industrial research seems to have rhetorical strength in these overly state-centred societies, but granted public control measures commercial research is also viewed as a viable path to medical progress.

There have been no similar studies conducted in Denmark. However, there is no particular reason to expect major national differences. The findings from the Swedish and Norwegian studies are reflected in British studies (Busby 2004; Haimes and Whong-Barr 2004), which indicate that the general trust in the authorities and willingness to support the progress of research without detailed contracting through elaborate informed consent procedures are not limited to Norway and Sweden alone. Furthermore, by the end of 2005 only seventy-six people had entered the Danish registry which was established in 2004 to ensure everybody the right to decline biobank research participation once and for all. Worth noting, in the same period the registry had fifty-two researchers registered as users of the registry: apparently, the right to withdraw tissue samples from research participation is a more significant issue to policy makers and researchers than to the general population.

The donors to Scandinavian biobanks count on the state to establish the necessary incentives and control systems for securing ethically sound biobank research. Their worry is that the incentives could change, and that research could be placed out of reach of the control systems. This is expressed in a typical way by a participant in the HUNT focus group study: ‘As we’re living in a well-organized society with laws and regulations and circulars, I’m pretty confident that [the samples] won’t be abused in the society we have at the present. And will continue to have – in my day, anyhow. It is however hard to tell anything about the society of the future, because it [the society] seems to be governed pretty much by economics, capital and those kinds of values.’ In this perspective, it is the appointed public institutions which should be trusted to exclude harmful and unnecessary research projects – not the individual donors. According to the researchers, there is a conflict between the imperative to utilize biobank research material, and the restrictions put on research by the consent and other requirements.

A paradox of the situation is that the lawmakers intended to empower the participants, while the thinking of the interviewed participants is that they expect the authorities to take responsibility. The political attempts to integrate the individual into evaluating research confuse the participants. To be positioned to evaluate the soundness of the research is seen as inadequate, which illustrates a basic contradiction of empowerment: ‘on the one hand “we” provide according to “their” needs, yet on the other hand “we” tell “them” their needs’ (Grace 1991: 339).

Rethinking the logic of biobank regulation

In view of the ambiguities of the regulation enacted in the respective countries; the complaints from researchers and health professionals; and the mismatch between the problems and solutions articulated by legislators and the concerns articulated by the donating public – the logic of biobank regulation must be rethought. It is necessary to ask: What is actually regulated with the laws and circulars on biobanking and why?

An official answer is that *biobank laws regulate researchers' access to human tissue through the consent requirement.* In this perspective, consent should ensure the negative freedom of the participants, in order to protect them from harm. Another possible answer is that *biobank laws are more about regulating the state/citizen relationship than about regulating research.* The handling of stored tissue by public authorities is important because it indicates the care exercised by the neo-liberal welfare state towards its citizens. In order for research to make use of biobank information in a legitimate way, the individuals taking part must waive their right to privacy, and exercise their self-determination by giving an informed consent. Looking at the legislation in the three jurisdictions, it appears that it is not enough to support research by coincidence: you must *want* to support research. If you participate, you approve of the research, otherwise you should decline participation. The overwhelming majority of people in the Scandinavian countries which participate are thus individually and actively linked to biobank research. The consequence of setting the individuals free is thus to link them closer to the research and to make them internalize the research aims. The informed consent requirement in a sense involves a micro-political mechanism: it makes it necessary for participants to actively engage with the agenda put forth by the authorities (Petersen 1997). They have to enact the values embedded in research participation.

How are notions of identity, privacy and control consequently shaped? Based on the exploration of the similarities above, we argue that the modus operandi of the regulatory efforts in the field of biobanking exemplifies a paradoxical enactment of citizenship: an ideal of individual autonomy is fused with a notion of the education of the masses; a search for the uncontaminated and independent citizen as arbiter of moral judgement is fused with strong efforts to enrol this 'subject' in public health policymaking. The participants partake in the responsibility for research making use of their contribution to a biobank – otherwise they should have opted out.

The activities exhibited in relation to regulation in the field of genetics include hearings and a constant call for more public debate. In these calls the authenticity of the voice of the individual seems to rest on this individual being 'uncontaminated' by knowledge about research, unpolluted by politics (Brown and Michael 2002). For a stance to acquire legitimacy it must represent the true voice of people; true autonomy. In a sense, the choice of the individual – to participate or abstain – comes to serve as the final arbiter of moral truth. Ironically, however, in hearings representatives of the public are then heavily educated by experts; in relation to donations they are heavily informed before allowed to decide on participation.

In a sense, the informed consent process becomes an education process. To consent without reading the consent form is not a 'proper' informed consent. In this way, to argue that studies show that participants do not find informed consent important is beside the point. The informed consent is an end in itself – it is part of the guide for a meeting of the citizens and science. Through consent procedures, biobanking becomes a mass education project. A suitable Millian narrative could be: due to prevailing paternalistic ideas, and the average level of education, the citizens have been treated like children by public health service of the past. Nowadays they might be treated like youth – but in the end they will be treated like adults: entrepreneurs of their own health, both in private prevention of disease and in consulting the health service.

Regulating research

The political agenda might present solutions which neither fits the worries of the researchers nor the participants, because the impetus for regulation has more to do with finding ways of ensuring trust, positive attitudes and legitimacy than with tightening the rules for biobank research. If current Scandinavian biobank regulation is aimed more at intervening in the state/citizen relation than at controlling research as such, then Julie Black (1998) might be touching important issues with her claim that the regulation presenting itself as ethics, generally serves to facilitate research rather than regulate it. This sustains our claim about the type of innovation that biobank laws represent: they re-delegate the interest of politicians to a re-enactment of the state contract. They confine legitimate interventions to ensuring negative entitlements, like the right not to participate in research. This might indicate a shift in emphasis in the Scandinavian welfare states from powerful but heavily obliged states to facilitating states.

What, then, regulates research? This is a complex question to answer. Some studies of the regulatory efforts in the field of genetic research present the ambition of regulation as failed (Wright 1994). Politicians try, but then succumb to pressures of financial competition etc. In contrast, Gottweis (1998) has argued that regulation indeed takes place, but it exhibits more complex modes of interaction between the fields of politics, economy and science. As Gottweis points out, there are numerous modes of regulation. Parliamentary regulation of biobank research does not necessarily relate only to the laws entitled 'Biobank Acts'. The research infrastructures of Scandinavian biomedical research have been subject to intense reorganization through legal means during the past decade. In Denmark a new university law has been passed to make universities operate more like companies. Property regimes have been amended to transfer property rights from government employed researchers to the institutions employing them. Also, in Norway any patentable innovation is now the property of the institution funding the relevant research. Every Norwegian university has consequently established their own Transfer

Technology Offices to make innovation a part of the basic sphere of activity. In Sweden, university employees still hold the property rights of their innovations. Swedish universities are nevertheless intensively encouraged to pursue the establishment of spin-off companies and commercial gain from public research and do in fact work towards these aims (Achen 2004).

In the Scandinavian countries, the terms of university funding have also been changed over the last years: it is now increasingly posed as a demand for receiving public funding that a private investor is willing to co-finance the research. Funding for doctoral programmes moves to public–private partnerships. Money is taken from basic research financing and transferred to funding programmes where commercial applicability and relevance are stated criteria for access to the funds. Even at the level of individual researchers, modes of regulation are set in place to increase the competition, to accelerate scientific production (Gottweis and Triendl 2006).

Conclusion

Regulation of the incentives for research – like university funding – has important implications for the concerns articulated by the donors interviewed in Norway and Sweden. However, they are hardly addressed by the explicit 'ethical' worries of the legislators when talking about biobanking. In framing the questions of concern as a matter of ethics, biobank regulation narrows these concerns down to a protection of every donor's right to individually scrutinize and withdraw from biobank research. The attention of the researchers and donors is focused instead at the aims and priorities of biobank research. In this perspective sound research is safeguarded by the society in establishing good incentives for and control of research projects. Public, not individual, control and deliberation are needed to address these issues (Jonas 1984).

The hopes and fears of the population enrolled in biobank research reflects the imagined community of the welfare states and health systems in which donations are made (Busby and Martin 2006; Haddow *et al.* 2007) but the legal regulation known as biobank laws echoes a contractual relation between individuals (Skolbekken *et al.* 2005). Informed consent might be vital to avoid abuse of research participants, but is easily overloaded if used in a paradoxical way to simultaneously educate donors and posit them to evaluate research projects. The way to ensure trust and advance sound research might as well be not to emphasize consent, but to re-emphasize the primary concern of participants and researchers (Petersen 2005b): the political issues involved in promoting biobank research which strengthen the egalitarian and altruistic ideals of Scandinavian public health service.

Note

1 All translations from Norwegian in this text were made by LØU & JAS.

References

Achen, T. (2004) 'Actors, issues and tendencies in Swedish biotechnology', in M. Häyrinen-Alestalo and E. Kallerud (eds), *Mediating Public Concern in Biotechnology. A map of sites and issues in Denmark, Finland, Norway and Sweden* (pp. 113–55). Oslo: NIFU.

Baumann, Z. (1995) *Life in Fragments*. Cambridge: Polity Press.

Black, J. (1998) 'Regulation as facilitation: negotiating the genetic revolution', *The Modern Law Review*, 61 (5): 621–60.

Brown, N. and Michael, M. (2002) 'From authority to authenticity: the changing governance of biotechnology', *Health, Risk and Society*, 4 (3): 259–72.

Busby, H. (2004) 'Blood donation for genetic research: what can we learn from donors?', in R. Tutton and O. Corrigan (eds), *The Gift: The Donation and Exploitation of Human Tissue in Research*. London: Routledge.

Busby, H. and Martin, P. (2006) 'Biobanks, national identity and imagined communities: the case of UK Biobank', *Science as Culture*, 15 (3): 237–51.

Foucault, M. (1979) *The History of Sexuality, Vol. 1: An Introduction*. London: Allen Lane.

Fuge (2006) Fuges nyhetsbrev nr. 2/2006, available online at www.fuge.no.

Furness, P. and Sullivan, R. (2004) 'The Human Tissue Bill. Criminal sanctions linked to opaque legislation threaten research', *BMJ*, 328: 533–4.

Gottweis, H. (1998) *Governing Molecules. The Discursive Politics of Genetic Engineering in Europe and the United States*. Cambridge, MA: MIT Press.

Gottweis, H. and Triendl, R. (2006) 'South Korean policy failure and the Hwang debacle', *Nature Biotechnology*, 24 (2): 141–3.

Grace, V. M. (1991) 'The marketing of empowerment and the construction of the health consumer: a critique of health promotion', *International Journal of Health Services*, 21 (2): 329–43.

Haddow, G., Laurie, G., Cunningham-Burley, S. and Hunter, K. G. (2007) 'Tackling community concerns about commercialisation and genetic research: a modest interdisciplinary proposal', *Social Science & Medicine*, 64: 272–82.

Haimes, E. and Whong-Barr, M. (2004). 'Competing perspectives on reasons for participation and non-participation in the North Cumbria Community Genetics Project', in R. Tutton and O. Corrigan (eds), *Genetic Databases: Socio-ethical Issues in the Collection and Use of DNA*. London: Routledge.

Hartlev, M. (2005) *Fortrolighed i Sundhedsretten – et Patientretligt Perspektiv*. København: Forlaget Thomson A/S.

Helgesson, G. (2003) 'A Swedish standard for information and consent procedures in biobank research', in M. Hansson and M. Levin (eds), *Biobanks as Resources for Health*. Uppsala: Uppsala University.

Higgs, P. (2005). 'Risk, governmentality and the reconceptualization of citizenship', in G. Scambler (ed.), *Medical Sociology*. London: Routledge.

Hoeyer, K. (2005). 'Studying ethics as policy: the naming and framing of moral problems in genetic research', *Current Anthropology*, 46: 71–90.

Hoeyer, K. (2003) '"Science is really needed – that's all I know". Informed consent and the non-verbal practices of collecting blood for genetic research in Sweden', *New Genetics and Society*, 22 (3): 229–44.

Hoeyer, K. and Lynöe, N. (2006) 'Motivating donors to genetic research? anthropological reasons to rethink the role of informed consent', *Medicine, Health Care and Philosophy*, 9: 13–23.

Hoeyer, K., Olofsson, B. O., Mörndal, T. and Lynöe, N. (2004) 'Informed consent and biobanks: a population-based study of attitudes towards tissue donation for genetic research', *Scandinavian Journal of Public Health*, 32: 224–9.

Hoeyer, K., Olofsson, B. O., Mörndal, T. and Lynöe, N. (2005) 'The ethics of research using biobanks: reason to question the importance attributed informed consent', *Archives of Internal Medicine*, 165: 97–100.

Irwin, A. (2004) 'Constructing the scientific citizen: science and democracy in the biosciences', *Public Understanding of Science*, 10: 1–18.

Irwin, A. (2006) 'The politics of talk: coming to terms with the "new" scientific governance', *Social Studies of Science*, 36 (2): 299–320.

Jonas, H. (1984) *The Imperative of Responsibility. In Search of an Ethics for the Technological Age*. Chicago, IL: University of Chicago Press.

Kettis-Lindblad, Å., Ring, L., Viberth, E. and Hansson, M. G. (2005) 'Genetic research and donation of tissue samples to biobanks. What do potential sample donors in the Swedish general public think?', *The European Journal of Public Health*, 1–8.

Koch, L. (2004) 'The meaning of eugenics: reflections on the government of genetic knowledge in the past and the present', *Science in Context*, 17 (3): 315–31.

Lane, J.-E. and Ersson, S. (1996) 'The Nordic countries. Contention, compromise and corporatism', in J. M. Colomer (ed.), *Political Institutions in Europe* (pp. 254–81). London: Routledge.

Lupton, D. (1995) *The Imperative of Health*. London: Sage Publications.

Mulkay, M. (1993) 'Rhetorics of hope and fear in the great embryo debate', *Social Studies of Science*, 23: 721–42.

NOU (2001) Norges Offentlige Utredninger 2001: 19: Biobanker. Innhenting, oppbevaring, bruk og destruksjon av humant biologisk materiale.

NOU (2005) Norges Offentlige Utredninger 2005: 1: God forskning – bedre helse. Lov om medisinsk forskning, som involverer mennesker, humant biologisk materiale og helseopplysninger (helseforskningsloven).

Pálsson, G. and Harðardóttir, K. (2002). 'For whom the cell tolls', *Current Anthropology*, 43 (2): 271–301.

Patologidatabank (2006). 'Patologidatabank – Opbygning og Funktion' (3 May), unpublished work.

Petersen, A. (1997) 'Risk, governance and the new public health', in A. Petersen and R. Bunton (eds), *Foucault, Health and Medicine*. London: Routledge.

Petersen, A. (2001) 'Biofantasies: genetics and medicine in the print news media', *Social Science and Medicine*, 52: 1255–68.

Petersen, A. (2003) 'Governmentality, critical scholarship, and the medical humanities', *Journal of the Medical Humanities*, 24 (3/4).

Petersen, A. (2005a) 'Biobanks: challenges for "ethics"', *Critical Public Health*, 15 (4): 303–10.

Petersen, A. (2005b) 'Securing our genetic health: engendering trust in UK Biobank', *Sociology of Health and Illness*, 27 (2): 271–92.

Pierson, C. and Giddens, A. (1998) *Conversations with Anthony Giddens: Making Sense of Modernity*. Palo Alto, CA: Stanford University Press.

RAND (2000) 'Handbook of human tissue soucres – a national resource of human tissue samples', RAND Monograph Report, NBAC.

Reg. Prop. (2001) 'Biobanker inom hälso – och sjukvården', 2001/02:44:4. White paper for the Swedish Biobank Act.

Rose, N. (1999) *Powers of Freedom. Reframing Political Thought.* Cambridge: Cambridge University Press.

Stegmayr, B. and Asplund, K. (2002). 'Informed consent for genetic research on blood stored for more than a decade: a population based study', *BMJ*, 325 (7365): 634–5.

Skolbekken, J.-A., Ursin, L., Solberg, B., Christensen, E. and Ytterhus, B. (2005) 'Not worth the paper it's written on? Informed consent and biobank research in a Norwegian context', *Critical Public Health*, 15 (4): 335–47.

Turner, L. (2003) 'The tyranny of "genethics"', *Nature Biotechnology*, 21 (11), 1282.

Tutton, R. (2004) 'Person, property and gift: exploring languages of tissue donation to biomedical research', in R. Tutton and O. Corrigan (eds), *Genetic Databases: Socio-ethical Issues in the Collection and Use of DNA.* London: Routledge.

Vedel, J. B. (2004) 'Arkiver og Fremtidsskabende Anordninger – Et Studie i Konstruktionen af Danske Biobanker. 2–85. 2004. Institut for Informations- og Medievidenskab. Århus Universitet. Unpublshed thesis/dissertation.

Vyberg, M., Bjerregaard, B., Bak, M., Gram, I. and Hvolris, H. (2005) 'Patologidatabanken. Dansk Selskab for Patologisk Anatomi og Cytologi', *Ugeskrift for Læger*, 167 (12–13): 1404–1.

Wendler, David (2002) 'What research with stored samples teaches us about research with human subjects', *Bioethics*, 16 (1): 33–4.

Wright, S. (1994) *Molecular Politics. Developing American and British Regulatory Policy for Genetic Engineering, 1972–1982.* Chicago, IL: University of Chicago Press.

12 Framing consent

The politics of 'engagement' in an Australian biobank project

Beverley McNamara and Alan Petersen

Recent biobank developments reveal the considerable expectations attached to human genetic research. The promises associated with genetic research must, however, be questioned as untangling the genetic and other contributions to many diseases will make accurate prediction difficult if not impossible and thus the prospect of new treatments unlikely (Holtzman and Marteau 2000). Despite the promised research outputs being questioned, authorities are developing policies and programmes on the assumption that genetic research *will* eventually deliver improved health and other benefits. Reflecting this optimism, in recent years, genetic epidemiology has been accorded a key role in public health's effort to disentangle the genetic, lifestyle and environmental contributions to disease (e.g. Khoury *et al.* 2000). In many countries, genetics and other biotechnology innovations are seen as having the potential to not only advance the health of individuals and populations but also to contribute to economic prosperity and national identity (Jasanoff 2005: 5–8). The linking of research projects to the enterprise of nation-building is evident in countries that are biotechnology leaders such as Britain, Germany and the United States (Jasanoff 2005), as well as newer entrants to the field such as India, Estonia, and Australia (see, e.g. Bhardwaj 2004; Kattel and Anton 2004). However, despite the efforts of supporters of biobanks to persuade publics of their potential health, economic and national benefits, establishing consent and legitimacy for new projects has not been without difficulties. A major problem confronting proponents of biobanks is that the genetic information that they hold is seen by many people as fundamentally different from other kinds of health and personal information thus requiring special consideration in relation to regulation. Public fears about the potential for the misuse of genetic information, including the infringement of privacy, commercial profiteering, and genetic discrimination in insurance and employment, revealed in surveys and other 'consultations', challenge scientists and policy makers to find ways to achieve consent for new genetic research projects. This problem is acute in the case of the new generation, 'prospective' biobanks, which tend to hold DNA and personal medical and genealogical information over a long period of time for *unspecified* research purposes. In recent years, the limitations of existing 'ethical' protocols and other regulatory devices for addressing the substantive issues arising from the collection, long-term storage and use of this kind of information has become increasingly evident (see, e.g. Tutton and Corrigan 2004; special issue on 'Biobanks; Challenges for "Ethics"' in *Critical Public Health*, 15 (4), 2005). These limitations have been of such concern that a new array of ethics and governance protocols has been developed in response to biobank initiatives. Of interest in the context of this paper is the degree to which these rest principally upon the notion of consent, facilitated by programmes of public consultation.

Analysing a range of published information pertaining to a major new biobank initiative, the Western Australian Genome Health Project (WAGHP), this chapter examines how consent and legitimacy for this project has been discursively framed during its pilot phase. Specifically, we examine how the project is portrayed in its official communications (e.g. reports, websites, memos), especially in arguments for its establishment, and how 'the public' is constructed in its consultation and 'engagement' strategies. Information was collected through various means including regular monitoring of official websites for the Western Australian Institute for Medical Research Laboratory for Genetic Epidemiology (WAIMR Genepi) from July 2004 and the Joondalup Family Health Study (JFHS), the first phase of the WAGHP, from January 2006. The Laboratory for Genetic Epidemiology First Biannual Report (July 2003 to July 2005) provided details of the background and planning of the WAGHP. Both the report and the JFHS website give references to a broad range of popular media coverage of the project. The question of how biobank projects are 'framed' by their proponents during their formative phase in their effort to engender consent, we argue, is likely to be crucial to their ultimate public acceptance and the trajectory of their subsequent development.

Throughout the chapter we ask what lessons may be learnt from the WAGHP experience about contemporary processes of governance, particularly as they pertain to other similar population-based genetic research projects. As has been noted elsewhere patients, consumers and the general public are as much part of the democratic regimes of governance in the genomics field as are scientists and policymakers (Gottweis 2005: 196). Consultation and consent processes used in public engagement strategies as described in this chapter however, serve as techniques of persuasion and regulation whereby public participation is defined within a limited domain and seemingly predefined agenda. In the context of the Western Australian biobank project, which would appear to closely parallel some other recent biobank initiatives, such as UK Biobank, citizens are recruited and socially enrolled to the project through appeals to citizenship and inclusiveness.

Biobanks' 'engagements'

Biobank projects provide an important site for examining the contemporary workings of power. As Jasanoff (2005) notes, biotechnology has played a key role in political life over the last three decades and is likely to continue to do so in the twenty-first century. Comparative studies of national and regional debates surrounding biobanks and other biotechnologies can help us identify and make sense of wider changes in political culture (Jasanoff 2005: 14). These include the fracturing of the authority of the nation-state and the supplanting of the 'old' politics of modernity with its focus on rationality, objectivity and universalism with the 'new' politics of pluralism, localism and aesthetic emphasis in lifestyle and taste (2005: 14). Nicholas Rose (1999) has documented the emergence of new configurations of power 'beyond the state' characterized by an emphasis on freedom, choice and autonomy in everyday life.

Mirroring the situation in the UK and some other countries, an ideal of autonomy underlies the contemporary discourse of engagement within the WA Genome Health Project. Citizens are 'consulted' and called upon to give their 'consent' to the project and, in the process, are ascribed responsibility for the decisions they make in regards to participation. Increasingly, the 'ethics and governance' of biobanks is seen as a matter of citizen deliberation, if projects are to achieve legitimacy and be successful. In the case of UK Biobank, for example, early 'consultations' sought to ascertain publics' and stakeholders' views on the 'ethics and governance framework' for the project (Petersen 2005; Petersen 2007). Consultation exercises tend to be stakeholder-oriented in that they typically include only those who have expressed an interest in the issue (Royal Commission on Environmental Pollution 1988). The discourse of consultation and consent reflects a model of governance that focuses on individuals' relationship to the state – as defined by their rights and obligations *qua* citizens – rather than a broader consideration of collective interests and regulatory processes that scrutinize, for example, issues of ownership, access and broad public benefit. Ethical reasoning or theory, as a form of deliberating about scientific progress, though struggling to denote the 'right and the wrong way to apply new knowledge', is alas not an 'inventive force in its own right' but is 'responsive to past scientific inventions' (Konrad 2005: 32). While the scientific inventions stand as 'objective' and 'truthful', ethical reasoning is seen as only able to help deliberate on the best possible way to facilitate the progress of science. The exclusion of science from public deliberations on biobank projects such as the WAGHP serves as a powerful means of narrowing debate on the substantive issues that these projects give rise to.

In short, we question to what degree the consultation and consent strategies currently employed in biobank projects such as the WAGHP provide citizens with opportunities to participate in ways that can transform policy and practice. Furthermore, we find that this process of limiting opportunities provides insight into the broader workings of power in contemporary societies. It appears that citizens' obligations are being extended 'beyond the obligations of informed citizenry, to obligations to participate in wide-scale genetic surveillance' (Kerr 2003: 47–8). Their ability to engage at a level of political action is shaped and curtailed by educative programmes provided by the proponents of the biobank projects. Citizens' concerns are anticipated and provided for within narrowly framed 'ethics and governance frameworks' that focus on the privacy of information and protection from discrimination. As responsible citizens who exercise their 'choice' to participate in the collection of genetic, health and lifestyle information, the participants of biobanks are implicitly drawn into the government of genetic risk. Their genetic citizenship is premised on acceptance of personal risk but an ignorance of the much broader uncertainties and risks associated with the long term storage and use of information about populations. This analysis of the deployment of 'public talk' in the establishment of consent for a current biobank project reveals much about the current state of science–society relations in modern societies (Irwin 2006), in particular the limiting of opportunities for citizen involvement in substantive debates on developments that are likely to significantly shape the future. An important challenge confronting scholars who are committed to the democratization of science and culture is how they may best contribute to the reframing of debates so as to help revitalize political culture and facilitate wide-ranging deliberation on proposed developments *before* they proceed.

In assessing the contemporary discourse of engagement in relation to the WAGHP, framing analysis proves useful. As Jasanoff (2005: 23) notes, this method of analysis presents a fruitful way of asking questions about 'how issues are framed for public action in democratic societies'. Framing may pertain to the 'proper' objects of research and boundaries of expertise, the legitimate actors, mechanisms and institutions for dealing with emergent problems, and what are seen to be the important issues for debate and deliberation. By framing issues in particular ways – by organizing certain facts, claims, and values, and ignoring others, claims-makers have the potential to shape public knowledge and potentially public policy (Miller and Riechert 2000: 45; Nisbet and Lewenstein 2002: 361). A specific framing may be achieved through the selective use of language, metaphors and rhetorical devices, which establish a framework of expectations so that individual issues and events assume meaning as public issues in need of action (Nelkin 1995: 72–3). Many factors are likely to influence how biobank issues are 'framed', including local histories and political institutional arrangements and dynamics, the operations of expert networks, and prevailing religious and cultural values. Further, different socio-cultural contexts provide different representational opportunities, devices and styles.

The Western Australian Genome Health Project (WAGHP)

As the data on the WAGHP illustrate, 'ethics and governance' is framed narrowly in terms of a particular, individualistic conception of citizenship. In both the public documents and in the 'community engagement' strategies of the WAGHP, individuals are envisaged as active participants in genetic research. 'Consent' presumes an 'informed' understanding of the project, its aims and likely outcomes. But this is also paralleled by a discourse of *partnership* and solidarity between researchers and 'community'. Both strategies work to position the consenting citizen in such a way that they are obliged to accept a level of responsibility for the risks that may be uncovered by the research. The risks perceived by the proposers of the biobank are anticipated and dealt with in the 'ethics and governance framework'. However, by focusing on privacy and informed consent in the ethical oversight of the project, other important questions regarding the ownership, use and control of the information collected in the database are obscured. A resulting language of citizenship is one where autonomy and individual rights are privileged over alternative participatory frameworks. Active involvement of citizens in the WAGHP biobank initiative is limited with the project framed in such a way as to privilege passive consent. Consent is assumed to entail commitment to the nation-building exercise of bringing the Western Australian initiative to international recognition. In ways reminiscent of the UK Biobank (Petersen 2005), Western Australians are asked to 'invest in the future' by accepting that the benefits of the WAGHP are long term and possibly more likely to benefit later generations rather than themselves.

Although at the international level, biobanks have been the subject of considerable scholarly investigation (Árnason *et al.* 2004; Petersen 2005; Tutton and Corrigan 2004), little critical work of this kind has been undertaken in Australia. This is not to say that the growth in genetic science has escaped the attention of the government and, to some extent, other interested public groups. A two-year comprehensive public inquiry into the protection of human genetic information commissioned by the federal Australian government highlighted a number of concerns related to the development of human genetic databases in Australia (ALRC/AHEC 2003; Weisbrot 2003). However, given that the inquiry was led by the Australian Law Reform Commission (ALRC) and the Australian Health Ethics Committee (AHEC) it is not surprising that the methodology for this section of the inquiry focused upon an examination of pre-existing regulations of human genetic databases and concerns about privacy and discrimination rather than on other substantive issues such as ownership and control. The development of the WAGHP needs to be seen in relation to the ALRC/AHEC inquiry as the WAGHP proposal has been developed with the expectation that once implemented the recommendations of the inquiry would address the ethical and legal issues raised by the plans to biobank hundreds of thousands of DNA samples for an unspecified time.

At the time of writing, the WAGHP is still in development, but several complex factors have underpinned its development to date. A desire for credibility in the international arena would appear to 'drive' the scientists involved to portray the Western Australian resources as unique, with the potential to 'position WA biomedical research as a world leader by capitalizing on the amazing resources' available in the state (McEvoy 2005). The international significance of the project is repeatedly emphasized in documents published to date. The project's proponents draw attention to the unique set of 'linked' databases on health, stretching back over many decades. Unlike Iceland, where legislation was enacted to allow for the development of the Health Sector Database (Ministry for Health and Social Security 2006), Western Australia has had a Data Linkage System (WADLS) in place since 1995 (described in Holman *et al.* 1999). Data linkage means that the files of information about individuals available in independent electronic databases are given unique identifiers and can be linked together for the purpose of analysis. The successful development of the WADLS (see www.population health.uwa.edu.au/welcome/research/dlu/linkage) to date is partly due to the relatively cohesive local research community and its close relationship with the government department of health. The relatively small and geographically isolated research community has also meant that projects using WADLS data are managed by researchers known to one another and often working in collaboration. An epidemiological discourse has developed around this 'unique resource' so that the population is presented as geographically isolated, out-bred, manageable in size, relatively stable with a very low net migration rate, and populated with historically large 'settler' families. The proposal for the WAGPH is further legitimated by its cost effectiveness and, like other biobanks elsewhere, is seen to promise significant new 'discoveries' in clinical and genetic epidemiology, pharmacogenetics and therapies. As an indication of the WAGHP proponents' commitment to international collaboration, and as a way of putting the WAGHP 'on the map', the WAGHP has joined the Public Population Project in Genomics Consortium (P3G 2006).

The role of community consultation in framing genetic citizenship

While multiple discourses exist in the establishment of any genetic database (Marsden 2004), in the case of the WAGHP there is a notable division between information prepared for 'the public', whether this be in the arena of community consultation or through media, and information disseminated to a professional audience of fellow researchers, as in the P3G Consortium website and in presentations at scientific meetings. In a climate where genetic databases are reliant on 'the trust of the public' (Bunton and Petersen 2005, p.21; Levitt and Weldon 2005) and upon the willingness of citizens to donate genetic and lifestyle information about themselves, such a division has implications for genetic citizenship and processes of governance. It has been suggested that ongoing public involvement in the oversight and review of regulations for genetic databases is essential to ensure 'active trust' (Stranger *et al.* 2005). However, while it is productive to engage citizens in reviews of regulatory regimes, it is more likely that revised forms of governance are needed that encourage an active form of citizenship (Turner 2001). Enabling 'active' participation, as opposed to passive consent, is a key challenge in the governance of biobanks, particularly when the language of citizenship is so narrowly conceived, as it is in the case of the WAGHP. It suggests citizen involvement in shaping the development of the project and perhaps offering substantial critique of its aims and processes; however, current consultative mechanisms allow little scope for this level of participation.

Strategies deployed in relation to the 'community consultation' of the Joondalup Family Health Study (JFHS 2006) are instructive in this regard. Promoted as a 'contemporary, community-based family health study', the JFHS is considered a pilot or trial for a broad based WAGHP. Joondalup is a northern suburb of Perth, the capital city of Western Australia. According to the public website (JFHS 2006) it was chosen because the suburb is believed by the researchers proposing the study to be representative of the WA population in regard to 'key health findings' and to have a 'strong sense of community', is regarded as 'visionary' and has a 'high proportion of young families'. As a means of establishing legitimacy the Joondalup study is compared to the long-running Western Australian Busselton study (see http://bsn.uwa.edu.au/) which the proposers of the study claim 'has led to improved health outcomes in a number of chronic diseases such as asthma and cardiovascular disease'. The JFHS study is not promoted as a project with the express aim of developing a genetic database as is the case with the UK Biobank and the terms 'biobank' and 'genetic database' are not used in any of the literature associated with the study, nor in the community consultation strategies used to date. However, those who consent to participate in the study are asked to volunteer information regarding lifestyle and diet, undergo various measures including lung and cardiovascular function and donate a blood sample, which would be stored for biochemical and genetic analysis in the future. Although there are plans to link this data to pre-existing information in the Western Australian Data Linkage System (WADLS) and to link it to the Family Connections Genealogical Project, which connects information about genetically related individuals, this information has not been provided to potential participants in the study. Implicit in this decision is the assumption that issues regarding the scientific protocol of the study are separate from the kinds of 'ethical' issues that may concern 'the public'. As Weldon (2004: 171) notes in relation to discussions surrounding the UK Biobank, separating ethical concerns from scientific issues assumes that scientific developments are immune from such concerns and as such do not require public input. Further, as she observes, this kind of thinking assumes that 'ethical considerations apply only at the level of consent'.

Like UK Biobank, which is financed in part by the UK's Department of Health, the WAHGP receives government support. In 2005 the Western Australian state government provided funds to the researchers proposing the JFHS study to develop a community consultation strategy. According to Professor Lyle Palmer, the Head of the Western Australian Institute for Medical Research Laboratory for Genetic Epidemiology (see www.genepi.waimr.uwa.edu.au) and study leader, the community engagement program, 'will be the most extensive outreach program ever conducted for any medical research study in the world' (JFHS 2006). Similar to UK Biobank publications, JFHS documents emphasize the inclusive nature of the project, its adherence to 'best practice', legal safeguards ('WA will become a world leader in the legal protections for genetic privacy'), and the adoption of oversight and governance mechanisms. A 'Briefing document', which accompanies the programme report, explains that 'the most important part of this Study will be forging a strong partnership with the whole community'. This language of partnership continues:

> We have, and will continue to spend, considerable time talking with City of Joondalup residents to ensure they fully understand and are in agreement with all aspects of the study. As a critical part of building this partnership, we take our commitment to protecting the rights and privacy of future participants in our study very seriously.
>
> (WAIMR 2006a: 1)

The principle means of 'community consultation' used in the strategy was a survey staged in two phases – a community survey which was distributed to the 7,500 residents in the Joondalup area and a community discussion forum followed by a deliberative survey given to the participants in the forum. The purpose of the deliberative survey was to gauge if people's opinions changed once they had the opportunity to discuss the project with the study team along with 'other experts and community members' (WAIMR 2006b: 5). In other words, the 'consultation' was conducted like a scientific intervention study, which allowed for assessment of views before and after the intervention. Sixteen per cent of the residents completed the survey and 109 attended the workshop and completed the deliberative survey.

The study has been reported as having 'astounding' community support with 85 per cent of the residents completing the survey stating they were 'very' or 'quite interested' in volunteering for the study (DNAge 2006). The conclusions in the report outlining the findings of both surveys note 'a very high latent interest in participation' in the proposed study. However, the report then goes on to acknowledge that 'some of the key groups to the study organizers – younger people and families – are also the groups least likely to be interested in participation' (WAIMR 2006b: 28). No mention is made of what this might mean for the likely success of the project nor is there discussion of how much confidence can be had in the conclusions given the 'relatively small' sample size for the deliberative survey (2006b: 26). Emphasis on the positives of the surveys and the omission of information that would allow readers to assess their validity serves to lend legitimacy to strategies that may follow in their wake. Given that the results from the survey are likely to be used for such legitimizing purposes it is worthwhile considering how the survey was framed.

It is evident from the survey final report that the purpose of the survey was to ensure strong community consent that was considered essential for the project's success. The deliberative methodology involved pre-testing the public's reception to the idea, introducing an 'intervention', in this case the opportunity to discuss the issue, and then post-testing to determine if public responses had changed. Regardless of the results of the study, the methodology used presupposes either passive consent or disapproval. The survey is worded in such a way as to request consent. With very little background information given about the study, Joondalup residents were asked to indicate the degree to which the project is important and their likelihood of participating. They were asked to select from a predetermined list of issues and to assess the degree to which these issues were important in determining their decision (for example, would 'getting free testing for health and medical conditions by expert clinicians during the initial study' attract them to the study, or would 'the opportunity to do something good for the wider community' motivate them to participate). The choice of issues regarding possible concerns that people may have about the study, not surprisingly, focus on privacy and regulation of access to personal data. This approach to surveying publics' views, by framing questions exploring a predefined range of topics, allows researchers to control the agenda of debate and avoid discussion of the substantive issues that may arise if a more open-ended research method was employed. Revealingly, the executive summary notes that 'There is considerable information in the study that may be of use in developing effective communications strategies to encourage participation' (2006b: 3). This suggests an intention to engineer consent, which sits oddly with the aforementioned aim of the project; namely 'forging a strong partnership with the whole community'.

The decision to release certain information and withhold other information is crucial to the conceptualization of the genetic citizen in the discourse surrounding the Joondalup Family Health Study. The study provides participants with the opportunity to receive 'relevant results' from their 'free and thorough health check', which is conducted by a 'team of expert, specialist doctors'. Relevant results are those which 'indicate a potentially significant health issue' though it is not clear what these may be. However, a clear distinction is made between non-genetic feedback and genetic feedback as participants are told they will not be given genetic information as the results are not considered a useful assessment of health risk for the individual. By contrast, non-genetic information about their health will be released so that individuals can take personal responsibility for future decisions which may impact adversely upon their health. Genetic information is pooled as a population-based resource open only to 'bona fide health researchers', but also uniquely identified to link with other information about individuals. In a now well-recognized model of genetic citizenship this discourse privileges both 'professional's entitlements' and the 'public's obligations' (Kerr 2003). While advances in genomics have the potential to shift the 'focus of regulatory regimes from group risk to *individual susceptibility*' (Rose 2001: 11), participants in the JFHS are told that researchers are not interested in single gene disorders, but in the gene–environment, gene–gene interactions believed to be implicated in a range of complex diseases. Nevertheless, the biobank serves as a repository of information about individuals where single gene disorders could be explored if the clinical gaze took a turn in that direction. The regulators of the biobank therefore are directly implicated in the 'politics of life itself' (Rose 2001) and play no small part in the shaping of genetic citizenship.

Narrowing the public discourse

While it is essential to be cognizant of the specific historical, socio-cultural and political context in which each biobank arises, themes emerging from this analysis that appear to have more general salience can be identified. As noted, the WAGHP has in many ways taken its model from UK Biobank, and this is particularly evident in the framing of issues considered relevant for public consultation. Public documents, mostly available through the website, draw heavily upon the positive aspects of the project and the potential benefits to the health of individuals and 'the public' in general. This is perhaps not surprising as Rose (2000: 67) has noted in relation to the changing relationships between the state, science and the market, the new genetics has to be 'sold' to diverse audiences that include lay publics. It is telling that a number of discursive techniques used in the JFHS project website are modelled closely upon both the UK Biobank and other genetic database or longitudinal health studies' websites. The adoption of highly regarded 'patrons' and supporters supposedly provides a degree of credibility to the projects, with the JFHS advertising Professor Fiona Wood (2005 Australian of the Year) as its patron and Professor Barry Marshall (Nobel Prize in Physiology or Medicine in 2005) as an advocate of the project. The media launch for the JFHS included along with the mayor and other local dignitaries, a popular footballer to provide a familiar and trustworthy face to 'the public'. The 'frequently asked questions' used on the JFHS website are not necessarily based upon the real enquiries of Joondalup residents but are based upon assumptions made about the project and on information from other sites, particularly an early version of UK Biobank's website (Petersen 2005: 287). A narrowing of public discourse is evident in the formulation of pre-packaged questions and answers, which makes it valid to ask: to what extent are citizens able to shape policies and decisions that will affect them?

It is debatable also whether potential participants in the WAGHP or its pilot study, the JFHS, are able to assess the nature and implications of these ventures when the framework of expectations is built upon a public discourse informed by the 'oversell' of local media. Local newspapers have run articles entitled 'WA could lead the world in biotechnology', 'WA to become the world's biggest genetics lab' and 'Grant puts WA at cutting edge' (WAIMR 2005). In an article in the major Western Australian newspaper (*The West Australian* 2005) that announced support for the study, concerns from health consumer groups that the study lacked detail and failed to adequately address privacy concerns were dismissed by Professor Lyle Palmer, the leader of the project, who answered this concern by stressing the 'overwhelming support' for the study indicated by the survey. Of course the surveys themselves are likely to play a role in the shaping of public discussion by helping to legitimate the study and also in marginalizing interest groups that may oppose the study or seek to clarify its parameters and regulatory framework (see Dietrich and Schibeci 2003). As with UK Biobank the scientific merit of the JFHS is stressed. However, unlike UK Biobank which provides few details about the nature of the benefits of the study or of similar studies, the JFHS does attempt to address this issue. Nevertheless, it is questionable how the examples given are relevant to the proposed study; for example, the Raine Study 'showed that repeated ultrasounds do not benefit pregnant women or their babies' and the Children of the 90s (ALSPAC) study 'suggests that premature babies are more prone to emotional and behavioural problems like ADHD in later childhood' (JFHS 2006).

The degree to which participants in biobanks are asked to place their trust in the researchers and regulators of these resources has been commented on elsewhere (Levitt and Weldon 2005; Bushby 2004). In the case of the JFHS the claim that public trust is in large supply, supposedly evidenced in the 'overwhelming support' for the study, is premised upon limited 'community consultation'. The lack of acknowledgement that there are multiple publics who may offer diverse and possibly conflicting views is an obvious omission in documentation of the JFHS to date. The question of how to involve the identified key groups (younger people and families) who have proved difficult to reach is not spelt out. Further, one would expect there would be some concern among the proposers of JFHS and the WAGHP projects that the research will be applicable and acceptable to the indigenous population and the broad range of migrant and refugee groups in Western Australia. Interestingly the fall-back position of only involving those who 'consent' is used to justify an amorphous participant group, which is hardly reflective of the diverse multicultural Western Australian society. There is an expectation that those who agree to participate will conveniently fit the homogenous genetic profile preferred in the creation of the genetic epidemiological resource. Yet, somewhat ironically, many of those who do not fit the profile have arguably the worst health of all Australians, with the mortality and morbidity statistics for Aboriginal people a constant source of concern and embarrassment for the Australian government (Australian Human Rights and Equal Opportunity Commission 2006).

Discussion and conclusion

The WAGHP has adopted a familiar model of 'public engagement' employing methods that have been extensively commented on and critiqued by social scientists on a number of grounds (e.g. Cronberg 1995; Joss 1995; Renn

1995). Mechanisms such as community surveys and workshops or forums, are portrayed by the project's proponents as an unproblematic means for gauging 'the publics'' views. However, as noted, consultation exercises are stakeholder-oriented in that they tend to represent the views of those groups who are interested enough in the issue to participate. In the case of the WAGHP, it is noteworthy that the project failed to 'engage' those very groups – young people and families – that are of particular interest to the project. Despite this, the project's proponents appear to have reflected little on why this may be so and especially on how their images of 'the public' have shaped the provision and communication of information and may influence people's responses (Wynne 2006). Surveys, forums and other 'engagement' mechanisms are not neutral tools of information gathering and exchange, but reflect assumptions about those who are to be 'engaged', including their levels of 'understanding' and scientific literacy and propensity to act on information. Like the UK Biobank workshops, the forum for the Joondalup study appears to have been carefully managed, with the study team and other experts on hand to explain the study's aims. Forum participants are conceived as 'empty vessels' who need to be 'topped up' with information provided by the experts, who are then 'tested' for their levels of understanding and support. This reveals adherence to the so-called deficit model of public understanding that assumes that any opposition or lack of engagement is due to 'the publics'' ignorance of the project and its benefits, which can be corrected through the changing of views via more or better information. As with UK Biobank, the validity and values of science and scientific debates about the project are mostly excluded from deliberation (Weldon 2004). There is little scope for those participating in the surveys or forums or for broader publics to offer assessment of the overall validity of the science and of the scientists' visions, and the longer-term economic, social and political implications of the project or of biobanking more generally. The 'hyping' of the benefits of the project, assisted by the use of local celebrity figures and a generally positive media coverage of related events has served to overshadow and marginalize dissenting voices and to add legitimacy to the project. The media itself was not used by the projects' proponents to foster debate about the various substantive issues raised by the project but rather to 'sell' the project to 'the public' whom, it is assumed, will be the ultimate beneficiaries.

The particular framing of 'public engagement' in the WAGHP thus far highlights a fundamental problem plaguing many large science projects of this kind; namely, the apparent inability of the experts and projects' funders to reflect upon the science–society relationship and their own suppositions (including 'knowledge deficits') about 'science' and 'the public'. There has been a failure to acknowledge how science and scientists are implicated in the governance of populations through guiding conduct in certain preferred directions and neglecting the significance of particular ways of thinking, knowing and acting. In particular, the value of local knowledge and perspectives, and of recognizing publics' ambivalences and differences of viewpoints about science and its applications tend to be overlooked in favour of emphasizing universal or generalizable knowledge, broad 'public benefits' of developments and consensus of viewpoints and goals. Scientific research, especially if health-related like the WAGHP, is assumed to be necessarily aligned with 'the public good' and problems, insofar as they arise, are seen as mostly resolvable and manageable through 'ethics and governance' protocols and 'consultative' mechanisms. 'Problems' such as public mistrust are increasingly constructed as risk factors and subject to regulation. The aim of 'consultation' is mostly oriented to establishing consensus and legitimacy for projects, rather than inviting debate, criticism and the voicing of concerns. Clearly, creating mechanisms for *substantial* citizen participation in relation to biobanks such as the WAGHP presents a significant challenge for proponents and managers of such collections. It implies a genuine openness in relation to addressing the concerns and criticisms that citizens may have about them which may unsettle established ways of doing science and even serve to impede the development of projects. However, by choosing to circumscribe and limit participation and debate and to engineer consent, proponents of biobanks run the risk of alienating publics and exacerbating existing concerns that publics may have about science and the adequacy of regulation. Overblown claims about 'benefits' and the underplaying of concerns and risks may lead publics and stakeholders to become skeptical about and oppose future similar projects. The case of physician resistance in relation to Iceland's Health Sector Database should provide scientists and policymakers with a salutary lesson on how a biobank project may be jeopardized when key parties are excluded from debates (see Pálsson, Chapter 3).

Notwithstanding the 'promise' of better health, preventive interventions and new designs in drugs, espoused in public documents and websites, many of the claims-makers themselves are now expressing concerns that the expectations associated with genetic epidemiology are not overplayed. A 'cautious' approach and 'an intensive period of hypothesis-free information collection' is proposed by researchers who must allay the criticisms of skeptics from within the scientific community (Davey Smith *et al.* 2005: 1495). However, this measured approach is confined to an expert debate and the public discourse continues to be framed in an optimistic and hopeful manner. As in the case of UK Biobank, the WAGHP's processes of 'public consultation' are seen as an essential stage in the acceptance and success of the project. As yet, proponents and managers of these projects seem oblivious to how their own frameworks of knowledge and practices may actually generate concern and mistrust rather than engender trust. In the case of UK Biobank, despite extensive efforts by its proponents to present a positive portrayal of the project, it has attracted substantial criticism from a number of quarters (Petersen 2005). Although in its early stages of development, the WAGHP is likely to do the same when the goal of engineering consent is the ultimate aim of 'consultation'. With issues of trust and consent very much at the forefront of public discussions about new technology developments, it is crucial that scientists and science policymakers involved in biobank projects reflect on their own 'knowledge deficits' about the science–society relationship thus far and on how in the future they may encourage publics to actively participate in and influence developments in ways that are meaningful and useful to them.

References

Australian Human Rights and Equal Opportunity Commission (2006) 'Indigenous health needs to be addressed with a holistic rights-based approach', available online at www.humanrights.gov.au/media_releases/2006/20_06.htm (accessed 20 May 2006).

Australian Law Reform Commission and Australian Health Ethics Committee (2003) *Essentially Yours: The Protection of Human Genetic Information in Australia*, Report 96. Sydney: Commonwealth of Australia. Available online at www.austlii.edu.au/au/other/alrc/publications/reports/96/ (accessed 20 May 2006).

Árnason, G., Nordal, S. and Árnason, V. (eds) (2004) *Blood and Data: Ethical, Legal and Social Aspects of Human Genetic Databases*. Reykjavík: University of Iceland Press and Centre for Ethics.

Avon Longitudinal Study of Parents and Children (ALSPAC). Available online at www.alspac.bristol.ac.uk/welcome/index.shtml (accessed 19 May 2006).

Bhardwaj, M. (2004) 'Rich databases and poor people: opportunities for developing countries', *TRAMES*, 8 (1/2): 90–105.

Bunton, R. and Petersen, A. (2005) 'Genetics and governance: an introduction', in R. Bunton and A. Petersen (eds), *Genetic Governance: Health, Risk and Ethics in the Biotech Era*. London and New York: Routledge.

Bushby, H. (2004) 'Blood donation for genetic research: what can we learn from donors' narratives?', in R. Tutton and O. Corrigan (eds), *Genetic Databases: Socio-ethical Issues in the Collection and Use of DNA*. London and New York: Routledge.

Cronberg, T. (1995) 'Do marginal voices shape technology?', in S. Joss and J. Durant (eds), *Public Participation in Science: The Role of Consensus Conferences in Europe*. London: Science Museum.

Davey Smith, G., Ebrahim, S., Lewis, S., Hansell, A.L., Palmer, L.J. and Burton, P.R. (2005) 'Genetic epidemiology and public health: hope, hype and future prospects', *Lancet*, 366: 1484–98.

Dietrich, D. and Schibeci, R. (2003) 'Beyond public perceptions of gene technology: community participation in public policy in Australia', *Public Understanding of Science*, 12: 381–401.

DNAge (2006) 'Family support study astounding'. Autumn Perth: Western Australian Institute of Medical Research.

Gottweiss, H. (2005) 'Emerging forms of governance in genomics and post-genomics: structures, trends, perspectives', in R. Bunton and A. Petersen (eds), *Genetic Governance: Health, Risk and Ethics in the Biotech Era*. London and New York: Routledge.

Holman C. D. J., Bass, A. J., Rouse, I. L. and Hobbs, M. S. T. (1999) 'Population-based linkage of health records in Western Australia: development of a health services research linked database', *Australian and New Zealand Journal of Public Health*, 23 (5): 453–9.

Holtzman, N. A. and Marteau, T. (2000) 'Will genetics revolutionize medicine?', *The New England Journal of Medicine*, 343 (2): 141–4.

Irwin, A. (2006) 'The politics of talk: coming to terms with the "new" scientific governance', *Social Studies of Science*, 36 (2): 299–320.

Jasanoff, S. (2005) *Designs on Nature: Science and Democracy in Europe and the United States*. Princeton, NJ, and Oxford: Princeton University Press.

Joondalup Family Health Study (JFHS). Available online at www.jfhs.org.au (accessed 20 May 2006).

Joss, S. (1995) 'Evaluating consensus conferences: necessity or luxury?', in S. Joss and J. Durant (eds), *Public Participation in Science: The Role of Consensus Conferences in Europe*. London: Science Museum.

Kattel, R. and Anton, R. (2004) 'The Estonian Genome Project and economic development', in *TRAMES*, 8 (1/2): 106–28.

Kerr, A. (2003) 'Genetics and citizenship', *Society*, September/October: 44–50.

Khoury, M., Burke, W. and Thomson, E. (2000) *Genetics and Public Health in the 21st Century: Using Genetic Information to Improve Health and Prevent Disease*. New York: Oxford University Press.

Konrad, M. (2005) *Narrating the New Predictive Genetics: Ethics, Ethnography and Science*. Cambridge: Cambridge University Press.

Levitt, M. and Weldon, S. (2005) 'A well placed trust? Public perceptions of the governance of DNA databases', *Critical Public Health*, 15 (4): 311–21.

McEvoy, R. (2005) 'The WA Genome Project: Why do we need it?', *WA Medical Forum* (September), available online at www.genepi.com.au/media (accessed 20 May 2006).

Marsden, W. (2004) 'Analyzing multiple discourses in the establishment of genetic databases', in G. Árnason, S. Nordal and V. Árnason (eds), *Blood and Data: Ethical, Legal and Social Aspects of Human Genetic Databases*. Reykjavík: University of Iceland Press and Centre for Ethics.

Miller, M. M. and Riechert, B. P. (2000) 'Interest group strategies and journalistic norms: news media framing of environmental issues', in S. Allan, B. Adam and C. Carter (eds), *Environmental Risks and the Media*. London and New York: Routledge.

Ministry of Health and Social Security (2006) *Act on a Health Sector Database No. 139/1998*, available online at http://eng.heilbrigdisraduneyti.is/laws-and-regulations/nr/659 (accessed 25 May 2006).

Nelkin, D. (1995) *Selling Science: How the Press Covers Science and Technology*, revised edition. New York: W. H. Freeman & Co.

Nisbet, M. C. and Lewenstein, B. V. (2002) 'Biotechnology and the American media: the policy process and the elite press, 1970 to 1999', *Science Communication*, 23 (4): 359–91.

Petersen, A. (2005) 'Securing our genetic health: engendering trust in UK Biobank', *Sociology of Health and Illness*, 27 (2): 271–92.

Petersen, A. (2007) 'Biobanks' "engagements": engendering trust or engineering consent?', *Genomics, Society and Policy*, 3 (1): 31–43.

Public Population Project in Genomics (P3G). Available online at www.p3g consortium.org/waghp.cfm (accessed 15 May 2006).

Renn, O., Webler, T. and Wiedmann, P. (1995) 'A need for discourse on citizen participation: objectives and structure of the book', in P. Renn, T. Webler and P. Wiedmann (eds), *Fairness and Competence in Citizen Participation: Evaluating Models of Environmental Discourse*. Dordrecht: Kluwer Academic Publishers.

Rose, H. (2000) 'Risk, trust and scepticism in the age of new genetics', in B. Adam, U. Beck and J. van Loon (eds), *The Risk Society and Beyond*. London: Sage.

Rose, N. (1999) *Powers of Freedom: Reframing Political Thought*. Cambridge: Cambridge University Press.

Rose, N. (2001) 'The politics of life itself', *Theory, Culture and Society*, 18 (6): 1–30.

Royal Commission on Environmental Pollution (1988) *21st Century: Setting Environmental Standards*. London: The Stationery Office. Cited in A. Irwin and M. Michael (2003) *Science, Social Theory and Public Knowledge*. Maidenhead: Open University Press.

Stranger, M., Chalmers, D. and Nicol, D. (2005) 'Capital, trust and consultation: databanks and regulation in Australia', *Critical Public Health*, 15 (4): 349–58.

Turner, B. (2001) 'The erosion of citizenship', *British Journal of Sociology*, 52: 189–209.

Tutton, R. and Corrigan, O. (eds) (2004) *Genetic Databases: Socio-ethical Issues in the Collection and Use of DNA*. London and New York: Routledge.

The Busselton Health Study. Available online at http://bsn.uwa.edu.au/ (accessed 20 May 2006).

The West Australian (2005) 'Survey receives health support for study of Joondalup families', 18 December.

Weldon, S. (2004) '"Public consent" or "scientific citizenship"? What counts as public participation in population-based DNA collections?', in R. Tutton, and O. Corrigan (eds), *Genetic Databases: Socio-ethical Issues in the Collection and Use of DNA*. London and New York: Routledge.

Weisbrot, D. (2003) 'The Australian joint inquiry into the protection of human genetic information', *New Genetics and Society*, 22 (1): 89–113.

Western Australian Data Linkage System (WADLS), University of Western Australia. Available online at www.meddent.uwa.edu.au/go/sph (accessed 20 May 2006).

Western Australian Institute of Medical Research (WAIMR) (2006a) *Briefing Document: Privacy and Confidentiality Protections for Participants in the Joondalup Family Health Study*. Perth: WAIMR. Available online at www.jfhs.org.au (accessed 20 May 2006).

Western Australian Institute of Medical Research (2006b) 'Joondalup Family Health Study: Report on the results from the initial community survey and from the community workshop deliberative survey'. Perth: WAIMR. Available online at www.jfhs.org.au (accessed 20 May 2006).

Western Australian Institute for Medical Research Laboratory for Genetic Epidemiology. Available at: www.wager.org.au (accessed 20 May 2006).

Western Australian Institute for Medical Research Laboratory for Genetic Epidemiology (2005) First Biannual Report (1 July 2003 to 1 July 2005). University of Western Australia.

Wynne, B. (2006) 'Public engagement as a means of restoring public trust in science: hitting the notes, but missing the music?', *Community Genetics*, 9: 211–20.

13 Governing through biobanks

Research populations in Israel[1]

Barbara Prainsack

Introduction: on biobanks and Israel

Since the Icelandic Health Sector Database Project emerged in the late 1990s (Pálsson 2007), biobanks have been a heavily debated topic both in mass and academic media. Besides the 'usual' concerns discussed in light of the emergence of a new medical technology, such as informed consent, ownership, and genetic privacy (see, for example, Meslin and Quaid 2004; Hoeyer 2004; Austin *et al.* 2003), some biobanks seem to have raised an additional sort of question: What is a population? Which categories can be used in order to delineate it (Tutton, Chapter 10)?[2] The Human Genome Diversity Project, originally proposed in 1991 (Cavalli-Sforza *et al.* 1991), which aimed to explore the genetic diversity of the human species, was practically put on hold by controversies over group consent and what constitutes 'groupness' (Reardon 2004; Juengst 1998).[3] The developers of the Icelandic Health Sector Database faced criticism with regard to their claim of the 'homogeneity' of the Icelandic population (see, for example, Árnason *et al.* 2000). Some authors have accused all 'population thinking' in the context of human population genetics as contributing to racism, or representing a new sort of racism (Gannett 2001; see also Collins and Mansoura 2001; Duster 2005). Others contend that the category of 'race' in genetic research, however defined, cannot be avoided entirely (Kalow 2001; Risch *et al.* 2002), and call for a careful definition and use (Sankar and Cho 2002; Daar and Singer 2005). Particularly large-scale population-genetic biobank projects which aim to 'represent' the DNA of a given population in a biobank (such as the Estonian Genome Project; see Eensaar, Chapter 4), but also smaller initiatives focusing on one segment of a population only, sometimes uncover certain tacit conventions which, when exposed, resuscitate unresolved conflicts about legitimate and ethically 'correct' ways of relating to skin-colour, ethnicity, and religious identities (Prainsack and Hashiloni-Dolev 2008).

Herbert Gottweis (Chapter 2) draws a distinction between governance *of* biobanks (which is affected through laws, regulations, and – formal or informal – ethical guidelines) and governance *through* biobanks. The latter concept refers to the ways in which rationalities and categories by which samples and information in a biobank are stored feed into the public sphere. Whereas the first part of this chapter provides an overview of the governance *of* biobanks in Israel by discussing their legal, religious, and cultural embedding, the focus will be on the governance of individuals and populations *through* biobanks, particularly through demographic categories. In Israel, the use of ethnic and religious categories in the organization and practices of biobanks is relatively unchallenged because they have a long tradition of structuring public space. Also, today, they fulfil an important function in public discourse and politics.

One of the things we can learn from the case studies discussed in this volume is that biobank projects are more likely to obtain public support and trust if the concepts and terminologies which materialize in biobank practices correspond with established narratives in a particular society. In other words, the extent to which categories used to delineate and define populations in a biobank 'fit' with collective identities and demographic categories in a given society is an important aspect of success or failure of biobank projects. Simultaneously, the use of certain demographic categories in medical research reinforces their power in the public sphere (on the notion of co-production, see Jasanoff 2004a; 2004b). This renders biobanks to be more than only 'topics' and 'problems' of governance (see Gottweis, Chapter 2) but active agents in the maintenance and reinforcement of identities and 'groupness'. Biobanks can 'preserve' distinctions between insiders and outsiders in the context of a particular society by (often literally) writing them in blood.

Methodology

The research process started with establishing a map of biobank projects in Israel (by online and literature searches and enquiring among physicians and medical researchers in Israel). In a second step, three biobanks were chosen for deeper analysis. The objective was to select case studies that would display as much variety regarding their research objectives, their governance structures, and their social and legal embedding as possible.

The first case, the DNA collection of IDgene Pharmaceuticals, was chosen because it was the most prominent biobank in Israel at the time when I conducted my research (2004–5). It was a research biobank that had been intensely discussed among researchers, bioethicists, and policymakers in the medical field.

The second project, the Dor Yeshorim (DY) premarital genetic testing initiative, was selected because in contrast to IDgene it had not received a lot of attention from either bioethicists or policy makers. At the time of my fieldwork in Israel, only a few of my interviewees (see below) knew exactly how DY operates. Furthermore, DY is a biobank solely for diagnostic purposes; DNA is not used for medical research. Like IDgene, DY limits its scope to one population sub-group only (one which is primarily defined according to religious criteria, namely strictly Orthodox Jews. However, DY primarily targets Ashkenazi communities (Jews from European descent, excluding Balkan countries and Italy; about 2½ million residents in Israel are Ashkenazi Jews, which amounts to less than 50 per cent of all Israeli Jews) because of the historically higher level of endogamy in Ashkenazi population and therefore the higher number of carriers of genetic diseases.

The third project, The National Laboratory of the Genetics of Israeli Populations (NLGIP), represents a research biobank which – in contrast to IDgene, which focused on Ashkenazi Jews exclusively – comprises the DNA of several 'isolated populations'. Again, in contrast to IDgene, which committed itself to enabling research into particular diseases, the NLGIP collects samples irrespective of whether or not the donor suffers from a particular condition (consequently, their samples are often used as control samples in genetic research). The categories used to delineate the research population for IDgene and the NLGIP, respectively, disclose intriguing contrasts and overlaps.

Once the three case studies were identified, I contacted the persons in charge of them, and scheduled interviews. I then used the snow-ball system to identify further individuals who were involved in biobanking from the perspectives of research, bioethics, law, and policy. During my final period of fieldwork in Israel, I focused on representatives of patient organizations and 'lay' citizens in order to assess how individuals who were not directly involved in biobanks or the regulation thereof conceived of the topic. In case of DY, I was unable to obtain permission to interview the individuals carrying out the genetic diagnosis but discussed aspects of the project with its founder over the phone. (For more details, see Prainsack and Siegal 2006). In sum, I conducted about twenty semi-structured narrative interviews (Weiss 1995) with bioethicists, policymakers, biobank staff, representatives of patient support organizations, and 'lay' people in the period between January 2004 and December 2005. Different questionnaires were used for different professional groups (such as patient organization representatives, molecular biologists/geneticists, lawyers, etc.) and different biobanks. Questionnaires explored the genesis of the particular biobank project, its aims and aspirations, obstacles on the way of its establishment (ethical, political, societal, technical, financial), and societal/legal/ethical responses to the project. I also discussed with my interviewees the perceived benefits and risks of large-scale DNA databases in general and in the Israeli context in particular; I asked them to assess the current situation of legislation pertaining to DNA databases; and I inquired whether they thought there was anything particular about the Israeli context in what they were doing. Drawing upon previous experience with interviewing inside a small community of experts where everybody knows each other (Prainsack 2006), I refrained from voice-recording interviews but I took extensive notes instead. Non-anonymized quotes were sent to the interviewees for authorization prior to publication. The entire research process was informed by principles of the Grounded

Theory approach (Glaser and Strauss 1967, Charmaz 1990). Data-analysis and the generation of theoretical concepts were intertwined and took place over the course of two years. Also insights obtained from various contexts of participatory observation (for example, attending social events with researchers) were used for the generation of research questions and categories. The research project also entailed an analysis of policy documents, literature (mainly academic publications drawing upon or describing the biobanks in question), as well as media coverage (daily newspapers and weekly/monthly magazines) both in Israel and abroad. This was done primarily for contextualization purposes and not intended to be one of the main outcomes of this study.

The governance of biobanks: regulations and narratives

In its Declaration of Independence dating back to May 1948, Israel commits itself to being both a Jewish and a democratic state. This commitment is an important factor in the contextualization of the different points of gravity within the field in which biobank projects emerge. While like in any democracy, the parliament – and/or authorities responsible to the parliament – are entrusted with legislating and regulating medical research and biotechnology, religious teachings play a role as well, albeit in a more complex and less obvious way: in order for new medical technologies to be widely accepted, they should be in accordance with (or at least not in stark contradiction to) major religious teachings.

In January 2005, the Ministry of Health issued regulations on 'the establishment and utilization of genetic samples banks'. Besides the definition of central terms (such as 'identified' and 'unidentified' samples, 'DNA-samples bank', etc), they require anybody who plans to establish such a collection to seek approval from the Helsinki Committee on Genetic Experiments on Human Beings within the Ministry of Health. The decision of this Supreme Helsinki Committee will depend on parameters such as the size and character of the prospective bank, its purpose, measures to protect the 'security' of the samples, and also upon whether or not the possibility 'of injuring a certain public or [ethnic] community'[4] exists. Particular attention is given to the issue of transferring samples from Israeli collections to those abroad. Such transfers need approval from the Supreme Helsinki Committee, and must not contain any names of individuals (Ministry of Health 2005).[5] It is important to note that these regulations apply only to biobanks for medical research, not to those which serve a purely diagnostic purpose (such as DY).

Already in 2002, the Bioethics Advisory Committee of the Israeli Academy of Sciences and Humanities (IASH) had published a report on 'Population-Based Large-Scale Collections of DNA Samples and Databases of Genetic Information'. The recommendation welcomed the establishments of DNA collections and pointed out the need to establish publicly funded population-based large-scale DNA collections (IASH 2002). Also here, emphasis was placed on both the prevention of the stigmatization of ethnic groups in Israel, as well as on preventing potential harm inflicted by transferring samples to other countries. It is apparent that regulators and bioethicists had been aware of the possible dangers of ethnic stigmatization in connection with biobanks early on. Whereas ethnic stigmatization and discrimination were recognized as a potential issue, the parameters according to which ethnicity is determined were not addressed in these documents.

Besides Judaism, which has already been mentioned as an important narrative shaping attitudes in the medical field, the heritage of Zionism – understood as the concept of a Jewish state in the Middle East operating according to European ideals of democracy, justice, and seculafrism – accounts for a large part of the considerable trust which people have in science and technology (Efron 2007). This is the case because science and technology have been conceptualized as crucial tools to enable the Jewish existence in the Middle East since the beginning of the Zionist movement. Without them, the promise to turn the barren desert into fertile land could not have been kept. Science and technology became symbols of the success of the Zionist project. The state itself has thereby obtained the role of a catalyst for modernization. 'Brainpower' is often praised as Israel's strongest resource in public speeches of politicians, and it is usually linked to high-tech. The opening of former President Moshe Katsav's speech on the occasion of Israel's Independence Day in May 2005 provides an illustrative example of the intimate relationship of the concept of independence with the tools that made it possible:

> Dear Friends, The people of Israel are celebrating 57 years of the renewal of Jewish independence and sovereignty with the establishment of the State of Israel. We are proud of the achievements of the State of Israel, which is one of the leading nations in the world in the fields of science, technology, medicine, and agriculture'.
>
> (Katsav 2005)

Today, science and technology have not ceased to be seen as important instruments to maintain, and now also to protect, the Israeli collective (Ben-Ari 2006; Golan 2004; Prainsack and Firestine 2005; 2006).

Besides the positive evaluation of science and technology, Zionist heritage also entails another aspect which is relevant for an understanding of the absence of a widespread problematization of ethnic categories: some authors argue that Zionism has always employed a biological understanding of the collective. Israeli geneticist Raphael Falk, for example, points out the important role that a biological understanding of collectivity played in the early history of Zionism:

> While most European Jews tried to fight against the idea of Judaism being a 'race', prominent Zionists such as Hess, Herzl, Bialik, Nordau

> and even Buber argued that the biological dimension of the Jewish 'Volk' should not be overlooked.
>
> (Hashiloni-Dolev 2004: 73; see Falk 2002)

Other authors also studied the ways in which 'Jewish scholars and scientists engaged in discourse about race and the "Jewish question" in order to counter anti-Semitism and to bolster either assimilationists or Zionist goals' (Kirsh 2003: 631; see also Efron 1993; Hart 1999). Kirsh contends that the Zionist movement adopted the concept of race from the national movements in Europe at the time of its inception (Kirsh 2003: 634).

How do Israeli biobanks relate to these patterns? A closer look at three biobank projects which cover a wide range of scientific objectives and vary with regard to purpose, practices and institutional structures, will help to answer this question.

Case study I: IDgene Pharmaceuticals[6]

IDgene had been an early dream of Ariel Darvasi, a biologist and computer scientist at Jerusalem's Hebrew University's Life Science Institute. Prior to founding IDgene, Darvasi had been associate director of human genetics and head of statistical genetics at SmithKline Beecham (now Glaxo SmithKline) in the UK. Upon returning to Israel in 1999, he started to put his idea of establishing a biobank into practice. His scientific approach, combined with the promise of access to a 'genetically homogenous' population, enabled him to obtain US$12 million in venture funds.

IDgene Pharmaceuticals Ltd aimed at the discovery of the genetic basis of common diseases (diabetes mellitus types I and II, multiple sclerosis, schizophrenia, Parkinson's, Alzheimer's, hypertension, asthma, breast cancer, and colon cancer) in Ashkenazi Jews, and at the development of drug targets and diagnostic markers. Methodologically, IDgene employed a population association-based strategy; this means that the analysis starts with identifying a candidate gene for a disease and then tests whether individuals affected by a certain disease have that particular gene (version or mutation). Since family clustering in DNA scanning was not used, genealogical records were not involved. Samples were collected through a network of fifty Israeli hospitals and 300 clinicians, with the informed consent of the donors. Until 2004, the company had collected 16,000 blood samples from individuals with four Ashkenazi Jewish grandparents (according to location of birth) suffering from the above mentioned diseases. The genetic profiles of the donors were then compared to the genetic profiles of a control group of Ashkenazi Jews who did not suffer from these diseases. IDgene's founder, Ariel Darvasi, had formulated the hope that diagnostic firms would want to buy non-exclusive licences to the technology platform in order to develop tools for identifying patients' genetic susceptibility to disease and identifying candidates for preventive therapy or closer monitoring. IDgene never aimed at developing drugs alone (interview Darvasi). Already by 2001, the company had collected about 10,000 samples from Ashkenazi donors; however, this had risen to 16,000 at the time of my first interview with Darvasi (spring 2004). The company's aim was to obtain 500–1,000 samples per disease, a number regarded reasonable for conducting association analysis.

Besides the fact that Israel is seen as ideal for this kind of research, because of 'its highly developed Western-style medicine and well-trained researchers' (Darvasi, quoted from [Judy] Siegel 2001), Darvasi identified as a main asset the genetic homogeneity of Ashkenazi Jews, who are 'even more homogenous than their Icelandic counterparts' (Friedlin 2000). Stanford University geneticist (now at the University of California, San Francisco) Neil Risch shared this view and argued that a further advantage of studying the Ashkenazi population was the large number of possible sample donors (Machlis 2001). Compared to Ashkenazi Jews, Risch stated, 'the Icelandic population has yet to demonstrate its utility' to be studied in the Icelandic Health Sector Database (quoted from [Joshua] Siegel 2001).

Risch thereby established Ashkenazi Jews as the perfect 'research population' in the field of genetics: They are portrayed as sufficiently large in number, and as sufficiently homogeneous, in order to deliver meaningful results.[7] In addition, due to the relatively high willingness of Israelis to donate blood for research,[8] Ashkenazi Jewish DNA donors are also relatively accessible, which Birenbaum Carmeli notes is 'facilitated by the participation of Jewish scientists that alleviates ethical concerns as well' (Birenbaum Carmeli 2004: 76). This latter aspect is important as any allusions to inhumane treatment in the past, when Jewish research subjects were not only examined but also killed by non-Jewish 'scientists', must be avoided if a research initiative is to be successful. The fact that IDgene focused on Ashkenazi Jews, who had been a target of the atrocities of Nazi 'medicine', had indeed led to concern among bioethicists during the phase of setting up the biobank. This concern, however, could eventually be appeased, because, as a molecular biologist put it, Ashkenazi Jews did not run a high risk of becoming a stigmatized group in Israel: 'Ahskenazi bashing is not very effective in this country. They are still the [social] elite' (Interview molecular biologist). In a context free from definitions and objectives imposed by outsiders, the ethnicity of Ashkenazi Jews can be naturalized without running the risk of stigmatization. This group has also traditionally dominated the top of the social and political hierarchy in Israel. The alleged homogeneity of Ashkenazi Jews, which outside of the context of safeness and security provided by the Jewish state could be considered as dangerous and harmful, can serve as a neutral scientific tool inside Israel: 'The Ashkenazi Jewish population is a resource that must be harvested quickly – before the "melting pot" effects of intergroup marriages in Israel result in the mixing of different ethnic subpopulations' (Zak *et al.* 2002: 441; see also Shohat 2000).

IDgene benefited not only from common understandings of Ashkenazi Jews being a particularly genetically homogenous group, but also from the widespread image of their pioneering spirit. As Rakefet Sela-Sheffy points out, the 'antagonism between the so-called "Oriental" [*Mizrahi*] and "Western" [*Ashkenazi*] identity' is closely linked to the 'tension between a secular modern national culture and a traditional one' (Sela-Sheffy 2004: 480). Ashkenazi Jews symbolize a secular and modern version of Israeli identity, the so-called *Tsabra* archetype: 'an ideal personification of the patriotic, collectivist and altruistic' Israeli ethos. In this context, it seems 'natural' that Ashekenazi Jews would be at the forefront of medical research, both as researchers, as well as DNA donors. What Steve Olson said about Israeli Jews in general is applicable to Ashkenazi Jews – the main sufferers from Nazi atrocities – in particular:

> If anyone has a right to be worried about the potential misuses of genetics research, the Jews do. But in conversations with Jewish friends and Israelis I've received quite another impression. People are proud of their history and want to know more about it. If that knowledge has medical benefits, all the better.
>
> (Olson 2002: 119)

It is apparent in the case of IDgene that the public understanding of who and what Ashkenazi Jews are, defined according to religious and ethnic categories, co-determined the structure of the biobank. Simultaneously, the alleged genetic homogeneity of this group emphasized in the context of research feeds back into the public sphere, where it reinforces the 'groupness' of Ashkenazi Jews.

Case study II: Dor Yeshorim – a databank for premarital genetic testing[9]

The Dor Yeshorim (DY, 'The Generation of the Righteous') initiative employs a unique policy of using genetic testing for the purpose of preventing courtship and marriage of two carriers of a recessive genetic disease (especially Tay Sachs). It is restricted to 'ultra'-Orthodox and Orthodox Jewish communities, where arranged marriages are common and where the prevalence of inheritable genetic diseases is probably even higher than in the general Ashkenazi population.[10] In addition, abortions are deemed permissible only in an extremely narrow range of cases in these communities. Founded in the early 1980s, DY operates today both in Israel and in the United States and has tested more than 200,000 men and women (Ekstein 2004).

Young individuals are tested in their early adulthood for certain genetic conditions – mostly recessive, fatal or severely debilitating conditions. The current list of screened diseases includes Tay-Sachs, Cystic Fibrosis, Gaucher's disease type I, Canavan disease, Familial Dysautonomia, Bloom syndrome, Fanconi anemia, Glycogen storage disease type 1A, Mucolipidosis type IV, and Niemann-Pick disease type A. It is worth noting that not all diseases enlisted are fatal, and the severity of the disease may vary among individuals. Gaucher's disease is illustrative of such variability, and in this case, testing is currently performed only on demand (see also Levene 2004). Tests are carried out mostly at educational institutes, namely Jewish high schools and Yeshivos. Consent is sought from parents in the case of minors under eighteen years of age. Testing fees amount to US$120–200 per person, while allegedly one half of the actual costs are subsidized by private donors and governmental support.

The tests' results are stored and individuals receive a confidential six digit ID number for future use. Names and other identifying details are disclosed by the blood donor. At the time of donation, tested persons may check their genetic compatibility with their potential spouse. Both individuals need to be in the database as no tests performed in other laboratories can be incorporated into DY's database. If both individuals are carriers of a particular recessive genetic disease, the match is deemed inadvisable. These 'incompatible' individuals are offered counselling. Counselling is not offered to all the thousands of other carriers within the database who choose not to marry another carrier for the same disease. The counselling itself is free, provided anonymously, and only by phone. Keeping in mind the early timing of the couple's genetic compatibility check, transpiring early in the courtship process or even before the couple has met, the damages of foregoing the option to marry are perceived to be minimal. Among the clients of DY, the vast majority of proposed marriages between carriers of the same genetic diseases are indeed cancelled.

While other premarital testing programmes (such as the thalassemia screening programme in Cyprus) (see Hoedemakers and ten Have 1998) provide information on the carrier status to tested individuals, DY only examines the 'genetic compatibility' of future spouses, and refrains from revealing any information on individual carrier status.

Historically, the emergence of the DY project is largely the result of the effort of one man, Rabbi Joseph Ekstein. Ekstein, an Orthodox Jew living in Brooklyn, New York, lost four of his children to Tay Sachs disease. At that time, in the late 1960s, Tay Sachs was a tantalizing problem in Orthodox communities. Carrier prevalence was high (Tay Sachs is very rare in the general population but potentially affects about one in every 2,500 Ashkenazi Jewish newborns); however, prenatal genetic testing was not a feasible solution because of the moral rejection of abortions among *Charedi* (so-called 'ultra'-Orthodox) populations[11]. Ekstein's search for allies in the fight against the disease was difficult at first. He recalls the experience of initial hesitance and resistance from other families within the community, who feared stigmatization. If it becomes known that one family member is a carrier, this might affect the marital chances of all other family members as well. 'Parents of sick children were afraid of their dirty laundry coming out in public' (quoted from *New Scientist* 2004). Initially, Ekstein had thought of solving the problem by introducing conventional genetic testing of young adults. He was soon convinced by community leaders and rabbis that stigmatization of disease carriers would render this practice harmful. He then invented the DY method, a screening system based upon premarital examination of the genetic compatibility of partners. According to Ekstein, '[s]ince Dor Yeshorim began, no genetically diseased children were born to the parents who used our tests' (quoted from *New Scientist* 2004).

Although not employed as a governmental programme, undergoing the DY test can be seen in some ways as socially mandatory: it is very difficult for strictly Orthodox couples to find a rabbi to marry them if they have not been tested (Chen 2001; Rosen 2003). This 'quasi-coercive' nature of the DY testing scheme, however, is mitigated by the fact that DY does represent a meaningful and relatively non-invasive way for young people who fear that they are carriers of genetic diseases to marry and procreate (see Prainsack and Siegal 2006). As a member of a youth webforum commented, 'You know you have to do it [undertake the test] eventually' (Prainsack and Siegal 2006). Not undergoing DY testing, in strictly Orthodox communities, is just not a 'rational' thing to do. The surprise expressed by non-Orthodox observers about the extent to which teenagers and young adults are willing to 'submit' themselves to the DY testing regimen underlines the large differences in practices, life-styles, values, and risk perceptions between *Charedi* communities on the one hand, and the majority population in Israel and other countries on the other. The main risk for a member of a strictly Orthodox community is typically not an infringement of his or her right to forego a genetic test, but rather the scenario of having an embryo/foetus affected with a lethal or life-threatening genetic disorder in the absence of the moral permissibility of abortion (see also Prainsack and Siegal 2008).

DY owes part of its success to operating within a segment of society which is largely separated from the rest of society (see Endnote 11). Members of strictly Orthodox communities adhere to norms and values of self-government that differ from those operating in the non- (or Modern-)Orthodox population. The same religious standards which make members of these communities abstain from intermarriage with the majority population also keep their DNA out of the sphere which is inhabited by the larger public in the reproductive field. The separation line between *Charedi* and other Jewish populations in Israel (see Sapir 2001) is mirrored by the (conceptually and physically) separate biobanks storing blood samples and information of Charedi and non-Charedi individuals, respectively, for the purpose of premarital and prenatal genetic testing.[12]

Case study III: the National Laboratory for the Genetics of Israeli Populations (NLGIP)

This collection, consisting of more than 2,000 immortalized human cell lines from individuals representing different populations in Israel, was established in 1994 as a joint initiative of the Israeli Academy for Sciences and Humanities and Tel Aviv University (TAU). Physically, the collection is located at the Sackler School of Medicine at the Tel Aviv University campus. The objective of the biobank is to facilitate research on 'complex diseases [. . .] and studies comparing disease-associated factors in different ethnic backgrounds', as they are, according to the director of the Laboratory, 'more likely to yield meaningful results when carried out in ethnically defined populations' (Gurwitz *et al.* 2003: 96).

The stored samples consist of DNA and cell lines derived from blood donated (with informed consent) by patients in clinics owned by two Israeli health funds. Blood is obtained only from donors who have four grandparents born at the same place (places of origin fall into one of eighteen categories, such as Ashkenazi Jews, North-African Jews (four subgroups), Oriental Jews (eight subgroups), Sephardic Jews (two subgroups), and Arabs (three subgroups). (For a comprehensive list, see Appendix.)

The idea of establishing the NLGIP emerged from a long-term collaboration of the founder and former head of the Laboratory, geneticist Batsheva Bonné-Tamir, with British geneticist Walter Bodmer. Bonné-Tamir had worked with the Samaritans and other isolated populations (Habbanites, Lybian Jews, Iraqi Jews, Moroccan Jews, Sinai Beduins, etc.) since the early 1960s (see Olson 2002: 114–19), and Bodmer had been interested in her work and visited Israel to obtain blood samples for his own research. 'Israel is a Garden of Eden for studying genetic diversity', Bonné-Tamir explains: 'We're a small country, but our people come from everywhere. That's why geneticists are so eager to work here' (quoted from Olson 2002: 110–11). It was Bodmer who had the idea of 'hav[ing] a laboratory of cell lines' first (interview Bonné-Tamir): when Bodmer was invited to give a lecture by the Israeli Academy of Sciences and Humanities, he presented this idea, which was welcomed immediately. The NLGIP at the Sackler School of Medicine became operational in 1994.

The Institutional Review Board (IRB) of Tel Aviv University had monitored the establishment of the NLGIP from its onset. It prohibited the offering of any monetary or other form of compensation to the donors of the blood samples, which stands in contrast to common practices of blood donation for healthcare purposes in Israel, where donors receive 'blood insurances'[13] for themselves and for their immediate relatives. In general, however, Bonné-Tamir explains that the cooperation between the NLGIP and the IRB has always been smooth (interview Bonné-Tamir).

The biobank staff makes considerable effort to convey the message that the Laboratory is a purely scientific endeavour, from which politics is absent. Bonné-Tamir refutes the accusation made by some (Israeli) bioethicists that some characteristics of the Laboratory's activities were 'racist'. 'I've been doing genetic and medical research in Israel since the 1960s', Bonné-Tamir states, 'and I've never been accused of scientific racism' (quoted from Olson 2002: 119). The current director of the Laboratory, David Gurwitz, explains that the use of homogenous ethnic groups is necessary for medical research:

'It's important that the control group is from the same genetic background [as the patients]. Most patients here are Jews and Arabs, not Austrians or Swiss' (interview Gurwitz). This quote alludes to several themes: First, it shows the explicitly medical context in which these population categories are mobilized. Second, it illustrates that the categories of 'Jews' on the one hand and 'Arabs' on the other, which the laboratory applies in correlation with their use in official governmental documents and in public discourse, function as the two main distinct demographic categories in the country. Third, it also hints at the particularities of the Israeli situation: The physical comfort and security in which most 'Austrians or Swiss' lead their lives stand in sharp contrast to the political 'state of emergency' in which many Israelis find themselves on a daily basis. This particular state of emergency is the lens through which the urgency of a problem or a task is evaluated. In this light, the main reason for the non-problematization of categories used to delineate populations in the medical research context in the Israeli public is the prevalence of more urgent topics and problems in the country.

As the word 'national' in its name indicates, the population groups represented in the NLGIP are seen in one way or another as those comprising the collective entity of 'Israelis' (even if not proportionally). In contrast to 'national'-labelled biobanks in other countries, which start with an abstract understanding of the population as a whole and then proceed to find criteria for stratification into sub-groups in order to achieve the highest possible representativeness of samples, the NLGIP was conceived the other way round. The starting point was not the abstract totality of the 'population', but the different sub-groups which constitute Israeli collective. While this can be explained by the academic biography of the NLGIP's founder, Professor Bonné-Tamir, whose work had focused on isolated populations, it resonates with the traditionally important role of collectivism and 'groupness' in Israel (immigrants arrived in groups from different countries and often settled and lived in this group-context for a long time; see also Ribke 2004: 144). On the other hand, it also reflects the political desire for clarity regarding the religious status and ethnic origin of Israeli citizens (see Joppke and Roshenhek 2001). For many scientists, Israel is destined to be a 'living laboratory' (Gurwitz *et al.* 2003: 96) for genetic research because:

> [b]eyond being so densely populated, Israel is distinctive in being exceptionally diverse ethnically, its residents comprise members of over twenty Jewish and Arab ethnic groups, who have kept their discrete cultural identities for many hundreds of years, and who have been exposed to minimal admixture through inter-marriages between the ethnic groups.
> (Gurwitz *et al.* 2003: 95)

The population categories that the NLGIP operates with are largely consistent with the ones used to categorize the ethnicity of citizens according to the Population Registry Law (1965) and the Identity Certificate Law (1982).

The latter established the duty for every Israeli resident aged sixteen or older to carry an identity card and present it on demand to a senior police officer, head of Municipal or Regional Authority, or a police officer of the armed forces on duty. Like in the NLGIP, the main demographic distinction in these laws is between 'Jews' and 'Arabs', with further differentiations according to geographical ancestry for Jews, and religious affiliation (Muslim, Christian, Druze) for Arabs. The category 'Israeli' is non-existent (since 2003, the 'nationality' [*leom*] category is no longer listed on Israeli identity cards).

Blood and identity

Both the ID cards in the political sphere, and the blood used for medical research have a linguistically intimate relationship to identity. The Hebrew term for identity card is *teudat zehut*. The same term, *zehut* ('identity') had occupied a central place in the public debate during the blood bank controversy in the late 1990s: when it had become public that Israeli health authorities had been secretly discarding blood donations from Ethiopian Israeli immigrants (out of fear of HIV contamination) for more than a decade, Ethiopian demonstrators expressed their anger and humiliation by reminding the public that 'the blood is our identity' (*ha-dam zeh zehut shelanu*; quoted from Seeman 1999: 175). By discarding their blood, representatives of the State had discarded the legitimacy of their identity, as a young man testified before a committee of inquiry: 'Throwing out our blood is like throwing out our identity' (quoted from Seeman 1999: 175). Moreover, by discarding their blood, authorities had also symbolically challenged the Jewishness of Ethiopian immigrants and citizens; a topic that had been contested from the very beginning of Ethiopian immigration to Israel. In religious Jewish teachings, 'the blood is the life/soul' (*ha-dam hu ha-nefesh*; Sefer Dvarim/ Deuteronomy 12: 23). A telling example of this was reflected in the words of an Ethiopian demonstrator after being tear-gassed: 'What are we, Arabs?' (quoted from Seeman 1999: 168).

An unintended consequence of the categories of ethnic and religious grouping employed by the biobank is that they provide a scientific basis for the 'groupness' of these ethnically and/or religiously defined collectives in public life. It is interesting to note, however, that in medical research involving Jewish subjects, homogeneity applies at two levels: First, it applies to Jews in general (see Carmeli Birenbaum 2004); second, it also applies to subgroups within the Jewish population. A difference between ethnic and religious grouping in the public sphere and ethnic grouping for medical research is that I have never encountered any implicit or explicit rank-ordering of population groups (Jewish and non-Jewish) in the context of Israeli biobanks. Whereas in the social sphere, the question to which ethnic and/or religious group one belongs is often an important indicator of both social worth and the 'quality' of Jewishness,[14] I never came across anybody in the field of medical research talking about one ethnic group being in any way better or worse, or more 'valuable' than the other. The only exception with regard to the (scientific) 'value' of an ethnically and/or religiously defined population group were scientists and policy makers referring to Ashkenazi Jews as more 'valuable' for research than others due to their alleged long history of endogamy and their large numerical size.

Governance through biobanks: protecting the borders

As Gottweis (Chapter 2) states, biobanks are more than mere objects of regulation. The Israeli case studies show that biobanks can play an important role in reinforcing collective identities. They can function as 'containers' which preserve the genetic components of the collective body. The collective body, an imagined entity comprising those who belong to a society or community (see Weiss 2002), serves as a point of reference when individuals act as citizens, as members of a state, and as members of a community. Images of the collective body also reinforce established notions of belonging.

The Israeli collective body, irrespective of the fact that about twenty to thirty per cent of Israeli citizens are non-Jews, is largely conceptualized as a Jewish body. This is closely linked to the particular circumstances under which the State of Israel came into being. Today, it manifests itself in the dominant position of Jewish heritage, religion and culture in the socio-economic distribution of wealth, education, and symbols, such as the Israeli anthem[15] and the flag with the Star of David. Whereas non-Jewish citizens enjoy the same legal rights as Jewish citizens, many claim that on a practical level, it is more difficult for non-Jews to obtain high political and economic positions (this is also due to army service, which is basically restricted to Jewish, Druze and Bedouine citizens,[16] being an important stepping stone into careers in business and the public sector) (see, for example, Peri 1983).

An interpretation of the Jewish character of the collective Israeli body as a manifestation of 'racism', however, would be rash. It must be seen in the context of a population which has suffered from centuries of persecution and murder and finally established its own state, from which the danger of annihilation has never disappeared. The current public debate about demographic threat – the scenario of the Jewish majority in Israel being outnumbered by non-Jews in the near future (see Prainsack and Firestine 2005; for skeptical portrayals, see Sussman 2004, Ettinger 2002, Judt 2003) – is a result of this feeling of danger. Not only military hazards but also the fear of becoming a minority in Israel itself and thereby losing grounds to justify the maintenance of the Jewish character of the state render the effective policing of the 'boundaries of the Jewish collective' (Kahn 2000: 72) as crucial. Maintaining clear distinctions between Jews and non-Jews, which plays out in population registries and the legal impossibility of cross-faith marriages[17] and adoptions, are labelled by some as a means to create an 'apartheid state' (Davis 1987; Glaser 2003), but are seen by others as crucial means for the survival of the Jewish collective in the Middle East.

By working with categories that carry strong references to family relations, genetic lineage, and religiosity, biobanks function as important repositories for these collective identities. On the one hand, the use of such operational categories is obviously induced by the objectives of genetic/genomic research, for which populations with limited genetic diversity are especially useful in many contexts: 'Despite some social scientists' criticism regarding the artificial nature of systems that are assumed to be closed, the targeting of population isolates for genetic studies is fairly routinized' (Birenbaum Carmeli 2004: 76).

Simultaneously, the non-problematization of these categories in the public sphere underlines the fundamental value they have for the identity and cohesion of the collective. Liquid nitrogen tanks for sample storage in biobanks can be viewed as instruments to guard and protect the borders of the collective and the cleavages within it. In two of the three case studies discussed in this paper, namely IDgene and the NLGIP, the genetic 'homogeneity' of the samples, in analogy to the supposed demographic homogeneity of the populations studied, plays a central role. This correlates with the traditional way of conceptualizing Jewish populations in Israel. As Kirsh has shown in her study on population genetics in Israel in the 1950s, 'Israeli researchers preferred to see the Jewish subject population as closed, unaffected genetically by non-Jewish neighbours' (Kirsh 2003: 646). If we skip the word 'genetically' in this quote, the intimate relationship between population categories in genetic/genomic research on the one hand and political and social categorizations of populations on the other is illustrated very clearly: both merge in the task of keeping the population 'unaffected' by non-Jewish neighbours, and past and future merge in light of the needs of the present. In this light, we can understand Bonné-Tamir's puzzlement over the accusation of 'racism' in the Laboratory as an expression of the deeper issue involved in the debate: What is at stake here is not ethnic purity, but the 'preservation', that is, the *continuity* of the existence of Jewish populations in Israel.

More generally, factors determining the failure or success of biobanks are often located in the social and political field more than in the field of science. Whereas the failure of a biobank project due to a flawed scientific design could have been expected, the actual reasons for crises of biobanks in recent years have been either related to problems of maintaining funding, and/or they were related to lack of trust and legitimacy of biobank practices (such as in Iceland, and the Human Genome Diversity Project). These practices cannot be understood independently from the social, religious, 'cultural', and political practices of the society they are embedded in.

Acknowledgements

The research for this paper has been funded by the GEN-AU (Genomeresearch in Austria) programme of the Austrian Federal Ministry of Science and Research, to which I express my sincere gratitude. I am also indebted to my interviewees and colleagues in Israel: Gil Siegal collaborated with me in researching and writing the part about the Dor Yeshorim initiative, and Rabbi Joseph Ekstein, its founder, commented on an earlier draft of the section on Dor Yeshorim. I am also grateful to Yechiel Bar-Ilan, Zvi Borochowitz, Stella Michelidou, Theresa Onell, and Alan Petersen for their valuable comments on the Dor Yeshorim section. I thank the editors and other contributors to this volume for stimulating comments and discussions. In addition, I thank Yael Hashiloni-Dolev, Ofer Firestine, Nurit Kirsh, Ursula Naue, Gísli Pálsson, Haran Rivlin, and Hendrik Wagenaar for helpful comments on the manuscript. The views expressed in this article are the views of the author, and mistakes remain solely mine.

Notes

1 This chapter is a revised version of an article titled 'Research populations: biobanks in Israel', published in *New Genetics and Society*, 26 (1): 85–103 (2007).
2 *Nature Genetics* devoted a supplement to the topic of '"Race" and the human genome', *Nature Genetics*, 36 (11), November 2004.
3 A preliminary goal of the Human Genome Diversity Project is to collect DNA samples from about 500 'distinct human populations' (see website at www.stanford.edu/group/morrinst/hgdp/faq.html#Q4) and turn them into 'the most complete worldwide human DNA collection that is available to not-for-profit researchers' (Cavalli-Sforza 2005: 335). The project also plans to 'carry out some basic, preliminary analyses of the DNA samples' (see website) primarily for the purpose of genetic diversity studies – but those might provide important spin-offs for medical research as well (see Cavalli-Sforza 2005). Besides conflicts over ethnic categories and groupness, the issue of patenting cell-lines also accounts for some of the opposition towards the project (Tauli-Corpuz 2001: 253; Amani and Coombe 2005).
4 The word used for community here is *kehilah*, which in this context applies to 'ethnic' communities in Israel.
5 Other important legal sources relevant for biobanking in Israel are the 'Patients Rights Law' 1996 (regulating informed consent, privacy protection, etc), the 'Genetic Privacy Law' 2000, the 'Law for the Protection of Privacy' 1981, and the 'Privacy Protection Regulations' 2001.
6 In 2004, IDgene Pharmaceuticals froze its activities due to financial difficulties. Ariel Darvasi still hopes to make his dream of identifying the genetic basis of common diseases come true.
7 See also Birenbaum Carmeli (2004: 76), who states that 'Jewish communities [. . .] are comparatively endogamous yet sizable'.
8 The prevalent positive attitude towards research among Israelis, and the willingness to contribute to medical research by donating blood, can be explained by a traditionally strong emphasis on healing and 'medical altruism' in Judaism (manifesting itself, for example, in the imperative that 'Thou shalt not stand idly by the blood of thy neighbor' (Leviticus 19: 16)). The design of the Israeli healthcare system according to the principles of solidarity and legality, in addition to the strong imprint that the social-democratic early history of the State has left on society in general, also helps to explain this phenomenon. In addition, as mentioned above, a positive view on the role of science and technology in society result partly from the Zionist heritage of the country.

9 For full papers on this topic, see Prainsack and Siegal (2006; 2008). Note that the 'correct' modern Hebrew spelling would be 'Dor Yesharim' (from the Hebrew word *yashar*, 'straight', 'righteous', 'upright'), however, we adhere to the Yiddish spelling, which is prevalent in strictly Orthodox communites.

10 The high incidence of genetically inheritable diseases among this group is said to be due to two phenomena: first, the 'founder effect', understood as the loss of genetic variation because of endogamy, and second, the so-called 'genetic drift', the inter-generational change of gene frequencies due to chance, instead of natural selection. Ashkenazi Jews are believed to descend from about 1,500 Jewish families dating back to the fourteenth century.

11 *Charedim* (pl. of the adjective *charedi*, 'fearful', or 'anxious'; more adequately translated as 'trembling in the awe of God'; see Isaiah 66: 2, 5) believe that *Halachah* (Jewish Law) should be observed literally and reject any progressive interpretation of it. Most *Charedim* live in tightly knit communities separated from the non-*Charedi* world. With respect to the use of technology, the rule is that whatever fosters the observance of commandments and does not conflict with other commandments or *Halachic* prohibitions will be accepted in *Charedi* communities.

12 It should be noted, however, that while DY is the most important institution in the field of pre-marital genetic testing in *Charedi* populations, it does not hold a monopoly. Some *Charedim* undergo genetic testing offered by other institutions catering to the needs of strictly Orthodox people. For more details, see Prainsack and Siegal (2006; 2008).

13 The 'Red Star of David' (*Magen David Adom* (MDA); the Israeli equivalent of the Red Cross) employs a blood insurance policy: When a person donates blood, he or she obtains 'blood insurance' coverage for him- or herself and close family members for the duration of one year. If during that year the donor or his or her family members are in need of a blood transplant, they will receive as much blood as they need. Of course, also 'uninsured' individuals receive blood transplants; however, they are responsible for finding people who afterwards donate on their behalf in order to compensate their 'blood debt' (my expression, BP).

14 For example, as Seeman (1999: 165) reports from demonstrations during the blood bank crisis, a young Jew of Ethiopian origin announced through a megaphone that '[w]e are as Jewish as the Yemenites, and more Jewish than the Russians!'.

15 The lyrics of *Ha-Tiqvah* ('The hope'), the Israeli anthem, include a section about the yearning of 'the Jewish soul' (*nefesh Yehudi*; for more information about the Israeli anthem, including an English translation of the first stanza, see www.science.co.il/Israel-Anthem.asp).

16 There is also a small number of Christian and Muslim Arabs who volunteer for army service.

17 Interfaith couples can bypass this obstacle by undergoing civil marriage abroad; the State of Israel then acknowledges their marital status.

References

Amani, B., and Coombe, R. J. (2005) 'The Human Genome Diversity Project: the politics of patents at the intersection of race, religion, and research ethics', *Law & Policy*, 27 (1): 152–88.

Árnason, E., Sigurgíslason, H. and Benedikz, E. (2000) 'Genetic homogeneity of Icelanders: fact or fiction?', *Nature Genetics*, 25: 373–4.

Austin, M. A., Harding, S. E. and McElroy, C. E. (2003) 'Monitoring ethical, legal, and social issues in developing population genetic databases', *Genetics in Medicine*, 5 (6): 451–7.

Ben-Ari, G. (2006) 'Innovation policy in the knowledge-based economy: the Israeli case', in E. G. Carayannis and D. F. Campbell (eds), *Knowledge Creation, Diffusion, and Use in Innovation Networks and Knowledge Clusters: A Comparative Systems Approach Across the United States, Europe and Asia* (pp. 253–82). Westport, CT: Praeger.

Birenbaum Carmeli, D. (2004) 'Prevalence of Jews as subjects in genetic research: figures, explanation, and potential implications', *American Journal of Medical Genetics*, 130A: 76–83.

Cavalli-Sforza, L. (2005) 'The Human Genome Diversity Project: past, present and future', *Nature Review Genetics*, 6 (April): 222–340.

Cavalli-Sforza, L. L., Wilson, A. C., Cantor, C. R., Cook-Degan, R. M. and King, M. C. (1991) 'Call for a worldwide survey of human genetic diversity: a vanishing opportunity for the Human Genome Project', *Genomics*, 11 (2): 490–1.

Charmaz, K. (1990) '"Discovering" chronic illness: using grounded theory', *Social Science & Medicine*, 30 (11): 1161–72.

Chen, S. (2001) 'The Rabbi said: marriage is forbidden (without genetic test)', *Yediot Aharonot* (in Hebrew), 19 August.

Collins, F. S. and Mansoura, M. K. (2001) 'The Human Genome Project: revealing the shared inheritance of all mankind', *Cancer*, 91 (suppl. 1) (1 January): 221–5.

Daar, A. S. and Singer, P. A. (2005) 'Pharmacogenetics and geographical ancestry: implications for drug development and global health', *Nature Review Genetics*, 6: 241–6.

Davis, U. (1987) *Israel: An Apartheid State*. London: Zed Books.

Duster, T. (2005) 'Enhanced: race and reification in science', *Science*, 307 (5712) (18 February): 1050–1.

Efron, J. M. (1994) *Defenders of the Race. Jewish Doctors and Race Science in Fin-de-siecle Europe*. New Haven, CT: Yale University Press.

Efron, N. J. (2007) *Judaism and Science: A Historical Introduction*. Westport, CT: Greenwood Press.

Ekstein, J. (2004) 'Responsibility of genetic testing', letter to the editor, *Genetics in Medicine*, 6 (5) (September/October): 465.

Ettinger, Y. (2002) 'The demographic threat – myth and reality', retrieved from www.acpr.org.il/cloakrm/opeds/020620.html.

Falk, R. (2002) 'Zionist settlement as a eugenic enterprise', *Alpayim*, 23: 179–99 (in Hebrew).

Friedlin, J. (2000) 'What's in a SNP? Israeli start-up searches Ashkenazi Jews for answers', *GenomeWeb*, retrieved from www.idegene.com/news/news_01.html (accessed 19 June 2003; website no longer active).

Gannett., L. (2001) 'Racism and human genome diversity research: the ethical limits of "population thinking"', *Philosophy of Science*, 63 (3) (suppl.): S479–92.

Glaser, B. G. and Strauss, A. L. (1967) *The Discovery of Grounded Theory: Strategies for Qualitative Research*. Chicago, IL: Aldine.

Glaser, D. J. (2003) 'Zionism and apartheid: a moral comparison', *Ethnic and Racial Studies*, 26 (3): 403–21.

Golan, T. (2004) 'Introduction', *Israel Studies: Science in Israel* (special issue), 9 (2): iv–viii.

Gurwitz, D., Kimchi, O. and Bonné-Tamir, B. (2003) 'The Israeli DNA and cell line collection: a human diversity repository', in B. M. Knoppers (ed.), *Populations and Genetics: Legal and Socio-Ethical Perspectives* (pp. 95–113). Leiden: Koninklijke Brill.

Hashiloni-Dolev, Y. (2004) 'Looking for the perfect child: fertility control and genetic management in Germany and Israel', Ph.D. thesis, Tel Aviv University, Israel.

Hart, M. B. (1999) 'Racial science, social science, and the politics of Jewish assimilation', *Isis*, 90: 268–97.

Hoedemaekers, R. and ten Have, H. (1998) 'Geneticization: the Cyprus paradigm', *Journal of Medicine and Philosophy*, 23 (3): 274–87.

Hoeyer, K. (2004) 'Informed consent and biobanks: a population-based study of attitudes towards tissue donation for genetic research', *Scandinavian Journal of Public Health*, 32 (3): 224–9.

Israel Science and Technology Homepage, information on the Israeli anthem. Available online at www.science.co.il/Israel-Anthem.asp (accessed February 2008).

Israeli Academy of Sciences and Humanities (IASH) (2002), 'Report of the Bioethics Advisory Committee of the Israeli Academy of Sciences and Humanities: population-based large-scale collections of DNA samples and databases of genetic information', December. Retrieved from www.academy.ac.il/bioethics/english/PDF/Finalized_Dna_Bank_Full.pdf.

Jasanoff, S. (2004a) 'The idiom of co-production', in S. Jasanoff (ed.), *States of Knowledge. The Co-production of Science and Social Order* (pp. 1–12). London: Routledge.

Jasanoff, S. (2004b) 'Ordering knowledge, ordering society', in S. Jasanoff (ed.), *States of Knowledge. The Co-production of Science and Social Order* (pp. 13–45). London: Routledge.

Joppke, C. and Roshenhek, Z. (2001) 'Ethnic-priority immigration in Israel and Germany: resilience versus demise', Working paper 45. San Diego: The Center for Comparative Immigration Studies, University of California. Retrieved from http://repositories.cdlib.org/cgi/viewcontent.cgi?article=1070&context=ccis.

Judt, T. (2003) 'Israel: the alternative', *The New York Review of Books*, 50 (16) (23 October).

Juengst, E. T. (1998) 'Groups as gatekeepers to genomic research: conceptually confusing, morally hazardous, and practically useless', *Kennedy Institute of Ethics Journal* 8 (2) (June): 183–200.

Kahn, S. M. (2000) *Reproducing Jews. A Cultural Account of Assisted Conception in Israel*. Durham, NC: Duke University Press.

Kalow, W. (2001) 'Pharmacogenetics, pharmacogenomics, and pharmacobiology', *Clinical Pharmacology and Therapeutics*, 70 (1): 1–4.

Katsav, M. (2005) Message from the President of the State of Israel Mr Moshe Katsav to the Jewish communities on the occasion of the State of Israel's 57th Independence Day. Retrieved from the website of the Israeli Ministry of Foreign Affairs, www.mfa.gov.il/MFA/MFAArchive/2000_2009/2005/President%20Katsav%20Independence%20Day%20Greeting%202005.

Kirsh, N. (2003) 'Population genetics in Israel in the 1950s: the unconscious internalization of ideology', *Isis*, 94: 631–55.

Levene, S. (2004) 'Genetic screening in the Ashkenazi Jewish population – is there a case for more testing?', Report for the National Screening Committee on the meeting held on 3 June 2004 at Guy's Hospital, London.

Machlis, A. (2001) 'Gene seekers look for big clues in a tiny population', *Financial Times* (25 June): 12.
Meslin, E. M. and Quaid, K. A. (2004) 'Ethical issues in the collection, storage, and research use of human biological materials', *Journal of Laboratory and Clinical Medicine*, 144 (5): 229–34.
(Israeli) Ministry of Health (2005) 'General Manager's publication regarding instructions of the Supreme Committee on Experiments on Human Beings regarding the establishment and utilization of genetic samples banks', No. 01/05 (2 January) (in Hebrew). An (unofficial) English translation can be obtained from the author (translation by Haran Rivlin).
The Morrison Institute for Population and Resource Studies, website of the Human Genome Diversity Project. Available online at www.stanford.edu/group/morrinst/hgdp/faq.html#Q4 (accessed February 2008).
New Scientist (2004) 'The Rabbi's dilemma' (Interview with Joseph Ekstein, by Alison George), 181 (2434) (14–20 February): 44–7.
Olson, S. (2002) *Mapping Human History: Genes, Race, and Our Common Origin*. Boston, MA: Mariner Book, Houghton Mifflin Company.
Pálsson, G. (2007) *Anthropology and the New Genetics*. Cambridge: Cambridge University Press.
Peri, Y. (1983) *Between Battles and Ballots: Israeli Military in Politics*. Cambridge and New York: Cambridge Middle East Library.
Prainsack, B. (2006) 'Negotiating life: the regulation of embryonic stem cell research and human cloning in Israel', *Social Studies of Science*, 36 (2) (April): 173–205.
Prainsack, B. and Firestine, O. (2005) 'Genetically modified survival: red and green biotechnology in Israel', *Science as Culture*, 14 (4): 355–72.
Prainsack, B. and Firestine, O. (2006) '"Science for survival": biotechnology regulation in Israel', *Science and Public Policy*, 33 (1): 33–46.
Prainsack, B. and Hashiloni-Dolev, Y. (2008) 'Religion and nationhood', in P. Atkinson, P. Glasner and M. Lock (eds), *Handbook of Genetics and Society: Mapping the New Genomic Era*. London: Routledge (forthcoming).
Prainsack, B. and Siegal, G. (2006) 'The rise of genetic couplehood? A comparative view of premarital genetic testing', *BioSocieties*, 1: 17–36.
Prainsack, B. and Siegal, G. (2008) 'Genetic testing and screening in religious groups – perspectives of Jewish Charedi communities', in D. Birenhaum-Carmeli and Y. S. Carmeli (eds), *Kin, Gene, Community: Reproductive Technology among Jewish Israelis*. Oxford and New York: Berghahn Books.
Reardon, J. (2004) *Race to the Finish: Identity and Governance in an Age of Genomics*. Princeton, NJ: Princeton University Press.
Ribke, N. (2004) '"The Ugly" vs. "The Beautiful" Israeli: struggles over defining Israeli identity', MA thesis, Tel Aviv University, Israel. Retrieved from www.tau.ac.il/tarbut/tezot/Ribke/Ribke-MA.pdf (in Hebrew; including summary in English).
Risch, N., Burchard, E., Ziv, E. and Tang, H. (2002) 'Categorization of humans in biomedical research: genes, race and disease', *Genome Biology*, 3 (7): 1–12.
Rosen, C. (2003) 'Eugenics – sacred and profane', *The New Atlantis* (summer): 79–89.
Sankar, P. and Cho, M. K. (2002) 'Toward a new vocabulary of human genetic variation', *Science*, 298 (15 November): 1337–8.
Sapir, G. (2001) 'Can an orthodox Jew participate in the public life of the state of Israel?', *Shofar*, 20 (2): 85–121.

Seeman, D. (1999) '"One people, one blood": public health, political violence, and HIV in an Ethiopian–Israeli setting', *Culture, Medicine and Psychiatry*, 23: 159–95.

Sela-Sheffy, R. (2004) '"What makes one an Israeli?" Negotiating identities in everyday representations of "Israeliness"', *Nations and Nationalism*, 10 (4): 479–97.

Shohat, M. (2000) 'The future of genetics: where are we going in the next forty years?', *Israel Medical Association Journal*, 2: 690–1.

Siegel, J. (2001) 'IDgene plumbs secrets of Ashkenazi gene pool. A population more useful for study than Icelanders?', *The Forward* (17 August), retrieved from www.idegene.com/news/news_13.htm (accessed 19 June 2003; website no longer active).

Siegel, J. (2001) 'Dawning of the age of genetic healing', *Jerusalem Post* (21 October), retrieved from www.jpost.com.

Sussman, G. (2004) 'The challenge to the two-state solution', *Middle East Report*, retrieved from www.globalpolicy.org/nations/sovereign/sover/emerg/2004/summer2 states.htm.

Tauli-Corpuz, V. (2001) 'Biotechnology and indigenous peoples', in B. Toker (ed.), *Redesigning Life? The Worldwide Challenge to Genetic Engineering* (pp. 252–70). Kingston, ON: McGill Queen's University Press.

Weiss, M. (2002) *The Chosen Body: The Politics of the Body in Israeli Society*. Palo Alto, CA: Stanford University Press.

Weiss, R. (1995) *Learning from Strangers. The Art and Method of Qualitative Interview Studies*. New York: Free Press.

Zak, N. B., Shifman, S., Shalom, A. and Darvasi, A. (2002) 'Genetic dissection of common diseases', *Israel Medical Association Journal*, 4: 438–43.

Appendix

Major ethnic groups represented in the Laboratory of the Genetics of Israeli Populations (see: http://nlgip.tau.ac.il):

A. Jews

Ashkenazi Jews	*Oriental Jews*	*Sephardic Jews*
	Iran	Bulgaria
North-African Jews	Iraq	Turkey
Algeria	Kochin (India)	
Libya	Kurdistan	
Morocco	Uzbekistan	
Tunis	Yemen	
	Ethiopia	
	Georgia	

B. Arabs

Bedouin
Druze
Palestinian

Index

eBooks – at www.eBookstore.tandf.co.uk

A library at your fingertips!

eBooks are electronic versions of printed books. You can store them on your PC/laptop or browse them online.

They have advantages for anyone needing rapid access to a wide variety of published, copyright information.

eBooks can help your research by enabling you to bookmark chapters, annotate text and use instant searches to find specific words or phrases. Several eBook files would fit on even a small laptop or PDA.

NEW: Save money by eSubscribing: cheap, online access to any eBook for as long as you need it.

Annual subscription packages

We now offer special low-cost bulk subscriptions to packages of eBooks in certain subject areas. These are available to libraries or to individuals.

For more information please contact webmaster.ebooks@tandf.co.uk

We're continually developing the eBook concept, so keep up to date by visiting the website.

www.eBookstore.tandf.co.uk

FSC
www.fsc.org
MIX
Papier | Fördert gute Waldnutzung
FSC® C083411

Kolibri 360 GmbH
Ferdinand-Jühlke-Straße 7
99095 Erfurt, Deutschland
produktsicherheit@kolibri360.de